GOOD DRUG REGULATORY PRACTICES

A Regulatory Affairs Quality Manual

Helene I. Dumitriu

Advisory Editor

Thomas O. Hintze

CRC Press
Taylor & Francis Group
Boca Raton London New York

CRC Press is an imprint of the
Taylor & Francis Group, an **informa** business

First published in 1998 by Interpharm Press, Inc.

This edition published in 2010 by Informa Healthcare

Published 2019 by CRC Press
Taylor & Francis Group
6000 Broken Sound Parkway NW, Suite 300
Boca Raton, FL 33487-2742

First issued in paperback 2019

ISBN 13: 978-0-367-44808-0 (pbk)
ISBN 13: 978-1-57491-051-3 (hbk)

Visit the Taylor & Francis Web site at
http://www.taylorandfrancis.com

and the CRC Press Web site at
http://www.crcpress.com

A CIP record for this book is available from the British Library.

Library of Congress Cataloging-in-Publication Data available on application

Contents

Part I
THE REGULATORY AFFAIRS QUALITY SYSTEM

Part II
THE REGULATORY AFFAIRS QUALITY MANUAL

Foreword

Awareness reveals that everybody active in today's business world depends on maintaining good relations with some kind of a customer or client. Thus, meeting the needs of our customers and clients has become the key to running a successful business. Quality is the essence of what the client or customer ultimately wants. Thus, we all are asked to deliver goods and services by focussing on these quality requirements.

In the pharmaceutical industry, well-established quality systems—GCPs, GLPs, and GMPs—exist; however, there is a narrow focus on specific functions or operations. Thus, a wide array of functions, although of utmost importance to the success of the overall business process—if not for the overall healthcare system—are only beginning to develop a sense for the quality aspects of their work. Regulatory Affairs is one such function that is starting to discover that quality, quality control, and quality assurance are essential elements for meeting the objectives of its operations. Regulatory Affairs is beginning to realize that a quality system designed to cover its specific work processes—and consequently named Good Drug Regulatory Practices (GDRPs)—will become a requirement that must be implemented in order to meet the challenges of the future. When asking for guidance and advice, however, little could be found, until the idea for this book began to materialize.

Helene Dumitriu, with 10 years of experience in Regulatory Affairs, provides comprehensive coverage of GDRPs as a truly holistic quality concept designed to meet the specific needs of Regulatory Affairs professionals. Starting with a theoretical basis, the author immediately turns to an organization's implementation of GDRPs as the true objective of this book. Therefore, a framework for assessing the quality of the existing operation of a pharmaceutical department is presented: A list of the key questions is compiled that needs to be answered during an initial audit—hopefully, all can be answered favorably within the organization. The Regulatory Affairs professional is not left alone, however, whenever a deficiency or a problem is discovered. Instead, the book provides suggestions and examples for improving the quality system by offering a series of policies geared to ensure dossier quality, timely submissions, and the maintenance of marketing authorizations. Policies and standards are presented in a generic fashion to allow for easy amendments and customizing for use in the environment of a specific company or organization.

To the best of my knowledge, this book is the first attempt to provide a holistic approach to GDRPs. Therefore, it deserves to be at the top of the essential reading list for all Regulatory Affairs professionals. Hopefully, this book will be of value to all in intimate contact with Regulatory Affairs, such as the company's project managers and quality assurance staff as well as those who work with Regulatory Bodies. In addition, management will find clues for improving and ensuring the quality of operations in its Regulatory Affairs function.

Dr. Thomas Hintze
Merck KGaA
July 1997

Preface

This book attempts to be the first systematic treatment on what should now be named Good Drug Regulatory Practice (GDRP). For many years, experienced Regulatory Affairs professionals have asked for a guide to the quality management of their work. Thus, this *Regulatory Affairs Quality Manual* was planned and written to meet this need. It is intended to serve as a reliable and useful source of information and background knowledge on the relationship between the quality of regulatory affairs work and the quality obtained and maintained of regulatory (e.g., marketing) authorizations. It will be a source of reference for Regulatory Affairs professionals as well as for quality controllers, who need information on the relationship between the work of Regulatory Affairs and quality in medicinal product development. However, this book does not contain information easily accessible elsewhere and/or likely to require frequent updates (e.g., addresses of Regulatory Bodies). The book covers regulatory affairs regarding global product development for the European Union, Japan, and the United States.

Part I presents a theoretical basis for quality management. Part II consists of policies and standards. Chapter 5 gives background information on policies. The policies (chapter 6) are written in a uniform format and cover the procedures and/or work results most frequently used or common in regulatory affairs. The information was assembled from a wide range of publications covering aspects of these important topics. It also relies on the author's extensive experience in over 10 years of involvement in international regulatory affairs, during which developmental and marketed medicinal products were dealt with on a worldwide basis. The policies cover, for example, Application for a Clinical

Trial License, Application for a Marketing Authorization, Education and Training, and Information Management. The standards include, for example, sample layouts of dossiers and the regulations that must be followed.

Acknowledgments

The author wishes to thank her advisory editor, Dr. Thomas Hintze, Merck KGaA, for his encouragement, valuable input, and helpful criticism of the draft. Without his continuous motivation, this work would never have been written.

Thanks also to Dr. Christian Spilles, Head Regulatory Affairs International, Bayer AG, for stimulating discussions.

Thanks also to Mrs. Gabriele Matuschek, who helped to transform the draft into a manuscript for publication.

Dr. Helene Dumitriu
July 1997

Introduction

A colleague once said: "Quality management ends with Good Clinical Practice [GCP]." He was right in that good practice has been extensively discussed for nonclinical and clinical development. However, Good Practice for Regulatory Affairs has not yet been defined.

This book was written to help improve this situation and to make Regulatory Affairs professionals, quality controllers, project managers, and senior management familiar with the contributions of Regulatory Affairs to the quality of medicinal product[1] development and the maintenance process. In order to have regulatory processes from which Regulatory Affairs professionals in both pharmaceutical companies and Regulatory Bodies can benefit, the concept of a holistic regulatory quality system is introduced. It should ideally comprise both Regulatory Bodies and Regulatory Affairs departments in the industry. In compliance with this idea, a Regulatory Affairs Quality Manual is proposed for the industry (with some suggestions for a complementary Quality Manual for Regulatory Bodies). It contains key questions to test the quality system, background information including, if applicable, regulatory requirements and measures for liability prevention, ready-to-use policies and standards, and points to consider during the development of the quality system.

The reader is invited to adapt the policies and/or standards to his or her organization and function and, if required, to develop quality assurance processes. These processes should form the basis of Standard Operating Procedures (SOPs).

---

1. Throughout this book, the term *medicinal product* will be used, rather than other synonyms used in literature (e.g., drug, drug product, pharmaceutical product, or proprietary medicinal product).

Part I

THE REGULATORY AFFAIRS QUALITY SYSTEM

1

Historical Overview

WHY REGULATORY AFFAIRS? THE DEVELOPMENT OF DRUG LAWS AND DRUG REGULATIONS

For an understanding of drug laws worldwide, some basic facts must be understood and borne in mind. Typically, drug laws are regulations after the fact—their development being triggered by unwanted and sometimes disastrous events. Information and knowledge on the use of medicinal products increase exponentially; thus, drug laws are proliferating at an ever growing pace. Drug laws provide for special Regulatory Bodies installed explicitly to ensure compliance with the drug laws. These bodies, once installed, begin to influence rule making according to their needs. The legislative efforts in the United States, Europe, and Japan have had a significant impact on pharmaceutical industry; therefore, the development of these legislations will be addressed in this chapter.

United States of America

Originally, medicinal products were not regulated by the federal government. Food and drug adulteration became a problem when the quality of products could no longer be guaranteed by the customer's knowledge of the producer and/or seller as a result of industrial scale production and mass transportation. When it was found that American troops had been supplied with substandard, imported medicinal products during the Mexican War, the first federal law specifically dealing with medicinal products was passed: the Import Drug Act of 1848. By 1901, the unsanitary conditions in the meat packing industry gave rise to public concern regarding the safety of food and, surprisingly, also of the quality of drugs. In fact, drug regulation may be considered as an outgrowth of food regulation. In 1902, distribution of contaminated diphtheria antitoxin in St. Louis caused the death of 12 children. This event forced legislators to pass the Biologics Act of 1902, which demanded the licensing of biological products and the production in licensed facilities. Between 1902 and 1907, the United States (U.S.) Bureau of Chemistry conducted studies on the safety of food additives with human volunteers that led to the passage of the Pure Food and Drugs Act of 1906 and, subsequently, the ban of dangerous food additives, such as borax, salicylic acid, formaldehyde, and copper sulfate. This first U.S. drug law prohibited the mislabeling and adulterating of medicinal products and introduced the U.S. Pharmacopeia (USP) and the National Formulary (NF) as official standards. The legal situation, however, still allowed for practices that from today's perspective appear somewhat bizarre. For example, in the case *U.S. v. Johnson,* the promoters of "Dr. Johnson's Mild Combination Treatment for Cancer" were charged for knowingly making false therapeutic claims.

The court's ruling led to the passage of the Sherley Amendment of 1912, which prohibits the labeling of medicinal products with false therapeutic claims. In 1937, the Massengill Company placed on the market a throat medicine that contained sulfanilamide dissolved in diethylene glycol, a common automobile antifreeze ingredient. No clinical trials had been conducted, as this was not mandatory at the time. One hundred seven people, mostly children, died after taking the medication. This tragedy forced legislators to react in a rigorous way: The Food, Drug, and Cosmetic Act (FD&C Act), which required proof of the safety of medicinal products, was passed in 1938. The industry was now allowed to place a medicinal product on the market 60 days after a New Drug Application (NDA) only if the Food and Drug Administration (FDA) posed no objections.

A further development is the Durham-Humphrey Amendments of 1951—also referred to as the "prescription drug amendments"—that divided products into over-the-counter (OTC) and prescription drugs requiring professional supervision. U.S. citizens were spared the thalidomide disaster, but the reaction of

Congress was the Kefauver-Harris Drug Amendments of 1962, thereby requiring proof of efficacy and formal FDA approval BEFORE placing medicinal products on the market, as well as introducing Good Manufacturing Practice (GMP) (1).

In the past 20 years, a number of major laws have been passed by Congress, such as the Orphan Drug Act, the Safe Medical Devices Act (SMDA) of 1990, the Medical Device Amendments of 1992, the Nutrition Labeling and Education Act of 1990, the Drug Price Competition and Patent Term Restoration Act of 1984 (Waxman-Hatch Act), the Prescription Drug Marketing Act (preventing illegal diversion and sale of prescription drugs), the Generic Drug Enforcement Act, and, most recently, the FDA Export Reform and Enhancement Act of 1996 (2). The movement toward reinventing the FDA has resulted in several FDA reform bills that are presently under consideration. Additionally to U.S. legislation, the FDA issues guidance documents that must be taken into account by industry. It must also be kept in mind that federal legislation applies to interstate traffic and that individual states have their own local laws.

European Union (EU)

The development of national drug legislation in the European Union can be demonstrated using Germany as an example (3). German legislation has its roots in the 19th century, focussing primarily on pharmacies. Originally, no marketing authorization for medicinal products produced on an industrial scale was required in Germany. When registration (i.e., notification of the product without submission of any proof of quality, safety, or efficacy!) was introduced in 1961 (4), this was done mainly to obtain an overview of the existing medicinal products: It turned out that 55,000 medicinal products were on the market at that time. (This large number is due to the fact that every strength and formulation of an active ingredient received a separate registration number.) The sleeping pill Contergan® (thalidomide) caused birth defects when taken by mothers in an early stage of pregnancy at the beginning of the 1960s: Children were born without arms—their hands starting at the shoulders; an estimated 10,000 children in Western Europe were affected. The appetite suppressant Menocil® (aminorex), registered in 1966, was responsible for several cases of death and was withdrawn from the market in 1968.

These events and the increased general perception of the risks to humans associated with the use of medicinal products for large-scale animal husbandry led to amendments in the legislation. However, the assessment of quality, safety, and efficacy by Regulatory Bodies prior to issuance of a marketing authorization became mandatory only when the European Community legislation (5) had to be introduced into national law. This led to the new German drug law of 1976 (6). In addition to introducing the requirement for a marketing authorization based on the assessment of the medicinal product's quality, safety, and efficacy, it also contained new regulations concerning pharmacovigilance for marketed medicinal products, the protection of clinical trial subjects, and liability.

The requirement to prove the efficacy of a medicinal product had been greatly debated at the time. It must be kept in mind that to this day many products, such as homeopathic and anthroposophical products, are marketed in Germany according to the principle of pluralism of scientific concepts. Major revisions were triggered in 1983 (7) because of misuse of drugs in food-producing animals, and in 1986 (8) by the experience report on the impact of the new drug law and the requirement to implement further Community legislation into national law (9). In the 1980s, the Alival® (nomifensin) scandal led to a revised alert procedure.

United Kingdom legislation can serve as another example for the development of national drug laws: Though attempts to regulate medicinal products date back to the Middle Ages, the simultaneous control of quality, safety, and efficacy is a very recent achievement and was introduced in the United Kingdom through the implementation of the Medicines Act of 1968 in 1971. This was a reaction both to EU legislation and to the thalidomide tragedy that had led to about 500 deformed children in the United Kingdom (10).

Today's national legislation in the EU Member States has lost its drive and basically mirrors the EU legislation, the development of which is described in the following.

European Community

The European Community (EC) was founded by the Treaty of Rome in the 1950s with the intention to establish a common market in order to "promote throughout the Community a harmonious development of economic activities, a continuous and balanced expansion, an increase in stability, an accelerated raising of the standard of living, and closer relations between the States belonging to it" (11). Significant amendments were introduced by the 1986 Single European Act, which implemented the concept of the Common Market—a market without any internal barriers—and reinforced the power of the European Parliament as well as of the EC institutions concerning environment, research, and technological development. The Maastricht Treaty signed in 1992 and enacted in 1993 fundamentally reshaped the European Economic Community into the European Union, which besides addressing common economic issues also includes a common foreign and security policy and cooperation concerning justice and security. A single currency is to be introduced and the responsibilities of the European Parliament were extended.

The EU legislation on medicinal products consists basically of Regulations, Directives, and Decisions. Regulations apply directly in all Member States. This instrument is typically used if the subject has not yet been regulated on a national level. Directives must be transferred into national law by the Member States prior to becoming effective. This instrument is used if there is a need to harmonize and adapt already existing national legislation. Decisions are measures intended to bind individual pharmaceutical manufacturers or Member States. The EU also publishes nonbinding recommendations and opinions to express the community view.

Marketing authorization procedures demanding the assessment of quality, safety, and efficacy of medicinal products were set first out in the Directive of 1965 (12). The multi-state procedure providing for mutual recognition of marketing authorizations by all involved Member States was established as well as the Committee for Proprietary Medicinal Products (CPMP), a body to facilitate the procedure (13). The multi-state procedure was later amended in order to make it more user-friendly (14). A special procedure, the concertation procedure, was introduced for high technology, especially biotechnology-derived medicinal products (15). Based on the experience with these two procedures, a new system installed in 1993 consisted mainly of the establishment of the European Agency for the Evaluation of Medicinal Products (EMEA) seated in London, the creation of a centralized procedure for biotechnology medicinal products (eligible also for other innovative products) leading to a binding Decision and a marketing authorization valid in all Member States, and the creation of a decentralized procedure for mutual recognition (16). These procedures will be reconsidered in 1998.

Additionally, the EU is issuing numerous Guidelines and Points to Consider documents that provide advice on requirements and procedures. A cornerstone is the Notice to Applicants of 1989 plus amendments (17) that describe the format for marketing authorization applications and procedures. Even though the EU guidance documents are not legally binding, they are supposed to reflect the state of the art and, therefore, are used by Regulatory Bodies as a basis for decision making. However, the high time of individual EU Guidelines appears to be over, even though the flood of advisory documents is ever rising. The single most important factor governing drug regulation today is the move toward harmonization between the EU, Japan, and the United States by the International Conference on Harmonisation of Technical Requirements for Registration of Pharmaceuticals for Human Use (ICH).

Japan

Legislation about medicinal products originally stems from the 19th century. Industrial production was first considered separate from that by pharmacists in 1961 when the Pharmaceutical Affairs Law of 1948 was divided into the present Pharmaceutical Affairs Law and the Pharmacists Law. The purpose was mainly to ensure the quality of medicinal products. However, after the thalidomide tragedy, regulatory focus turned also to safety and efficacy. Manufacturing and marketing authorization procedures were specified in 1967. Good Manufacturing Practice (GMP) was established in 1974 (year of enforcement: 1975), Good Laboratory Practice (GLP) in 1982 (year of enforcement: 1983), and Good Clinical Practice (GCP) in 1989 (year of enforcement: 1990). Because of liability suits such as the SMON (subacute myelo-optical neuropathy) case, the Adverse Drug Reaction Suffering Relief and Research Promotion Fund Law (currently, the Investigation Organization for Side Effects Relief and Research Promotion Law) was enacted in 1979. Provisions for orphan drugs and research were made by the revision of the Pharmaceutical Affairs Law in 1993 (18).

International Conference on Harmonisation (ICH)

Today, harmonization is a major factor in the development of drug laws in the EU, Japan, and the United States. Based on previous harmonization efforts, the ICH began to take shape in 1989 when the Steering Committee was established. Under its full, official name, "The International Conference on Harmonisation of Technical Requirements for Registration of Pharmaceuticals for Human Use", the ICH is a project of both the Regulatory Bodies and the pharmaceutical industry from the EU, Japan, and the United States. Its goal is to expedite the development and approval processes of medicinal products without sacrificing safeguards on quality, safety, or efficacy. Its sponsors are as follows:

- Commission of the European Communities (CEC) and European Federation of Pharmaceutical Industries' Associations (EFPIA)
- Ministry of Health and Welfare (MHW) and Japan Pharmaceutical Manufacturers Association (JPMA)
- Food and Drug Administration (FDA) and Pharmaceutical Research and Manufacturers of America (PhRMA)

The International Federation of Pharmaceutical Manufacturers Associations (IFPMA) is also an official participant and provides the ICH Secretariat. (Observers, such as Canada and the World Health Organization [WHO], may also be involved.)

The ICH serves as a forum for the identification and discussion of differences in the technical (i.e., scientific) requirements for marketing authorizations in the EU, Japan, and the United States. It makes recommendations for modifications of technical requirements in order to enhance mutual recognition, provide for more economical use of resources (human, animal, material), and suggest ways to achieve greater harmonization in the interpretation and application of technical requirements. Topics for harmonization are decided by the Steering Committee. Consultation is in a stepwise process:

Step 1: A preliminary draft is generated by relevant Expert Working Groups and, provided consensus has been reached, forwarded to the Steering Committee.

Step 2: The draft is transmitted to the CEC, the MHW, and the FDA for consultation according to their usual procedures. (In the EU, the CEC consults with the

EFPIA, which in turn distributes for discussion to national industry associations, who involve their members.)

Step 3: The revised draft is generated by the designated regulatory Rapporteur based on the comments received and then referred to the Expert Working Group for sign-off by the experts before referral to the Steering Committee for adoption.

Step 4: The final draft is endorsed by the Steering Committee and recommended for adoption in the EU, Japan, and the United States.

Step 5: The recommendations are transferred to the national/regional regulations or legislation according to national procedures.

Key events in the process are major meetings numbered consecutively: ICH1 took place in Brussels in 1991 (19), ICH2 in Orlando in 1993 (20), and ICH3 in Yokohama in 1995 (21).

World Health Organization

Another influencing factor for drug laws is certainly to be seen in the activities of the WHO. The WHO is an intergovernmental organization of 166 Member States within the Charter of the United Nations with the goal of all people attaining the best possible level of health. Its constitution came into force on 7 April 1948, commemorated each year as World Health Day. The WHO works through three principal bodies: the World Health Assembly, the Executive Board, and the Secretariat. It is a decentralized organization with headquarters in Geneva, Switzerland, and six regions—Africa, the Americas, Eastern Mediterranean, Europe, Southeast Asia, and Western Pacific—each with its own Regional Committee and Regional Office.

The WHO is the directing and coordinating authority on international health work and encourages technical cooperation for health with Member States. With regard to the pharmaceutical industry, important developments were the 1951 International Sanitary Regulations, renamed in 1969 as the International Health Regulations. WHO activities include support for ministries of health concerning the development of methods for assessing quality, effectiveness, and efficiency. Recently, the WHO has been paying special attention to primary healthcare implementation. WHO publications on medicinal products cover, for example, essential drugs, drug policies, drug regulation, the international pharmacopoeia, quality control, ethical guidelines, safety assessment, drug research and development, and laboratories.

Summary

In addition to legislation, the development of which has been outlined above, Regulatory Bodies have issued and continue to produce documents intended to provide advice to the industry on procedures and requirements. These documents, though not legally binding however, are used as a basis for decision making by the Regulatory Bodies, as they are supposed to reflect the current state of the art. The growing number of such guidance documents may be seen with some concern by the industry, as they may result in increasing development costs. Indeed, there is some danger of requirements augmenting unnecessarily by "Regulatory Creep".

In contrast to the marketing of many other products—and just like with cars and airplanes—the worldwide legal environment prohibits the marketing of a medicinal product unless the sale is explicitly allowed by an approved marketing authorization. That is why Regulatory Bodies have been created. It is a common misunderstanding that their job is to register your products. It is not. Regulatory Bodies have been established to keep medicinal products off the market unless the applicant can prove quality, efficacy, and safety.

REGULATORY AFFAIRS: DEVELOPMENT IN THE PHARMACEUTICAL INDUSTRY (22)

Why are Regulatory Affairs departments needed? The answer is simple: Because there are regulators and regulations.

As the regulatory environment has evolved continuously and has become extremely complex, (almost all) pharmaceutical companies have established Regulatory Affairs departments. They did so simply because companies need to understand and fulfill regulators' needs. Within internationally operating companies, national and corporate functions are often differentiated. In addition, to cope with EU procedures and the flood of Regulations, Guidelines, recommendations, and draft guidance documents, European liaison and Regulatory Intelligence functions have been created.

Regulatory Affairs is usually recognized as having three basic functions today:

1.   The outlet of the company to Regulatory Bodies

2.   The interpreter of regulations to companies

3.   The influencer of new regulations

All three make Regulatory Affairs THE interface between companies and Regulatory Bodies and, therefore, a key player in the medicinal product development and maintenance process. Although regulatory requirements were likely to be somehow met by nonexperts in the field of regulatory affairs in the past, today's rapidly evolving regulatory environment cannot be adequately coped with, unless you have dedicated Regulatory Affairs professionals.

REGULATORY AFFAIRS PROFESSIONAL SOCIETIES

With the developing regulatory environment and the parallel maturation of the regulatory affairs function in the pharmaceutical industry, a need was perceived to provide forums for discussion of relevant issues and to establish professional societies that would be distinct from trade associations.

In the past 20 years, regulatory affairs has evolved from a clerical activity to play a strategic role in medicinal product development, requiring broad scientific knowledge, regulatory expertise, and, in particular, great personal communication skills. Today's focus is on strengthening the identity of the profession as well as to

- Educate people who want to work in the regulatory affairs area.

- Confer on these people the designation of professionalism.

- Provide continuous training for Regulatory Affairs professionals.

Some major professional societies are described in the following sections.

The British Institute of Regulatory Affairs (BIRA) (23)

BIRA was founded in 1978. Persons professionally involved in regulatory affairs may become members—professionals from, for example, the pharmaceutical, cosmetic, toiletry, herbicide, pesticide, agrochemical, veterinary, animal health, and food industries. Though the majority of members are in the United Kingdom, there are also members from the rest of Europe, Australia, Japan, and the United States.

The objectives of BIRA are as follows:

- Establish professional identity and standards for Regulatory Affairs professionals.

- Promote education and science in regulatory affairs.

- Advance the professional competence of members and promote cooperative relations with other allied organizations.

- Collect and circulate relevant statistics and information.

There are standing committees that represent particular sectors (e.g., biotechnology, medical devices) and regional groups (e.g., BIRA East and BIRA North). BIRA offers meetings, trainings, and a diploma course as well as studies to gain a Master of Science degree in Regulatory Affairs (organized by the BIRA and validated by the University of Wales). BIRA publishes the *BIRA News* and the *BIRA Journal.*

The European Society of Regulatory Affairs (ESRA) (24)

ESRA was established by BIRA in 1986 to reflect the developing harmonization within the EU and to serve the needs of the profession in Europe. It is a society for professionals interested in European regulatory affairs (human and animal healthcare). It is based in London (its premises and Secretariat are shared with the BIRA, and many membership benefits are common between the two organizations). Its mission is to increase awareness of the importance of regulatory affairs in ensuring public health and the economic viability of companies in the healthcare sector. Its objectives are as follows:

- Information service (e.g., by publishing the *ESRA Rapporteur*).

- Education for members.

- Develop/promote best practices in regulatory affairs.

- Educate about all aspects of regulatory affairs as a profession.

- Liaise with other professional groups, media, and Regulatory Bodies.

Central European Society for Regulatory Affairs (MEGRA—Mitteleuropaeische Gesellschaft fuer regulatorische Angelegenheiten e.V.)

MEGRA was founded in 1988 and focusses on the German-speaking countries in Central Europe. It is an association of Regulatory Affairs specialists working for the pharmaceutical industry, Regulatory Bodies, or other organizations in the fields of medicinal products, devices, and diagnostics. Its goal is the advancement of regulatory affairs by education and specific training measures as well as by information on regulatory affairs in Germany, Austria, and Switzerland.

The Pan-European Federation of Regulatory Affairs Societies (PEFRAS)

PEFRAS is dedicated to improving the coordination and services offered by societies of Regulatory Affairs professionals in the fields of information and education.

The Regulatory Affairs Professionals Society (RAPS)

RAPS is an international society of Regulatory Affairs professionals working in healthcare (medicinal products, devices, biologics) systems regardless of their affiliation (government, industry, academia, or consultancy). It is committed to serving the professional development needs of the members by

- Facilitating the exchange of ideas
- Fostering cooperation among regulatory professionals in industry and government
- Providing continuing education in regulatory affairs

Founded in the United States in 1976, RAPS became an international society in 1985 and presently has a worldwide membership of over 6000 (25). RAPS is organized regionally, even if the U.S. chapter still has a dominating role today. RAPS Europe, created in 1985, focusses on regulatory activities within the EU, the European Free Trade Association (EFTA), and the Eastern European countries. The European educational program started in 1992. In 1995, the European Resource Center was installed, which provides an independent source of information for regulatory developments in Europe (26). Cooperative efforts include the European Commission, Ministries of Health in major European countries, the European Agency for the Evaluation of Medicinal Products (EMEA), Notified Bodies, as well as the FDA. RAPS also works together with national regulatory organizations and leading trade associations.

As a logical consequence of the general trend toward regulation of the profession, some academic institutions have started to offer postgraduate programs in regulatory affairs (27):

- Temple University, Pennsylvania, U.S.: Masters of Science in Drug Regulatory Affairs (requires 10 courses and the submission of a thesis)

- Long Island University, New York, U.S.: Masters of Science in pharmaceutics (with emphasis on regulatory affairs and quality assurance; requires 10 courses and the submission of a thesis)

GOOD PRACTICE AND QUALITY MANAGEMENT: AN OVERVIEW

In addition to drug legislation, good practice concepts and, lately, quality system concepts began to evolve as it was realized that the setting of requirements was not enough to guarantee quality medicinal products.

Good Manufacturing Practice (28)

The concept of GMP was established when it was realized that quality is determined by the manufacturing process rather than by subsequent quality control. GMP regulations aim to ensure the pharmaceutical quality of medicinal products and, therefore, regulate manufacturing personnel, facilities and equipment, documentation, manufacture, quality control, contract manufacture, product complaints, recall procedures, and self-inspections. The development of GMP was influenced mainly by the FDA, the WHO, the Pharmaceutical Inspection Convention (PIC), and the European Economic Community (EEC).

The term *Good Manufacturing Practice* was first used in 1962 in the Kefauver-Harris amendment to the Food and Drug Act in the United States; the FDA has set the pace for the global development of GMP. In the 1970s and 1980s, GMP became the subject of regulations in most countries. In 1969, the WHO published GMP guidelines. The PIC, established in 1970 by the EFTA, issued a GMP guide based on the WHO document. In Japan, GMP was established in 1974 and enforced in 1975. The EU published GMP guidelines in January 1989, which served as basis for the new PIC GMP guidelines of 1989 and the 1992 edition of the WHO GMP guidelines. In 1985, the Association of Southeast Asian Nations (ASEAN)—Brunei, Indonesia, Malaysia, the Philippines, Singapore, and Thailand—published the ASEAN GMP Guide. The present WHO GMP guidelines are based on the EEC and the ASEAN guidelines and are strongly influenced by the ISO 9000 series issued by the International Organisation for Standardisation (ISO) (see below).

Good Laboratory Practice

GLP originated (29) in the mid-1970s in the United States, when a variety of deficiencies were revealed in contract laboratories during an FDA inspection. These deficiencies included, among others, the conduct of studies, record keeping, archiving, and animal husbandry. The scandal widened when inspections of other laboratories revealed similar discrepancies. As a response, the chemical industry suggested the implementation of a quality management system, which was later named GLP. The FDA adopted GLP regulations in 1979. The Organisation for Economic Cooperation and Development (OECD) issued GLP guidelines in 1981 for chemicals. There is Community-wide legislation on GLP in the EU. In Japan, GLP was established in 1982 and enforced in 1983.

Good Clinical Practice (GCP) (30)

Cases of violations of human rights and scientific misconduct or fraud in connection with clinical trials in the United States in the 1960s and 1970s led the FDA to issue regulations and to control compliance. The European approach was characterized by a greater emphasis on guidelines and quality assurance rather than on quality control. The development of GCP in the EU was influenced by various initiatives, for example, by proposals of individual pharmaceutical companies, but also by FDA regulations, which in case of noncompliance would function as a trade barrier. A CPMP note for guidance on GCP was issued in 1990 and came into force in July 1991. The legal status was that of recommendations. Real legal power was given to GCP by Directive 91/507/EEC.

Another contribution to the development of GCP came in the form of the Nordic Good Clinical Trial Practice (GCTP) guideline of 1989. Between 1985 and 1990, several national GCP documents were published from various countries, including Germany, France, Italy, and Spain. Japan introduced GCP in 1989 (enforced in 1990). The early 1990s saw the rise of two global GCP initiatives driven by the WHO and by the ICH. The ICH has recently issued a Step 4 document that has been approved by the CPMP and will be mandatory for studies in the EU commencing after January 1997 (31). Yet another important influencing factor for GCP is the Declaration of Helsinki and the development of the Ethics Committee system.

ISO 9000 Series

ISO 9000 originates from the British standard BS 5750 governing military supply. Because of its special nature, BS 5750 concentrated on the manufacturing procedures and the quality systems of suppliers. The norms for quality systems were subsequently developed as the present international ISO 9000 series on a national and international level—North Atlantic Treaty Organization (NATO)—in order to be applicable to companies in general (even if best suited for manufacturers).

Whereas GMP aims to ensure the pharmaceutical quality of a product, aspects such as the overall organization of a company are not subject to GMP, even though the quality of the company's medicinal products and services may be significantly at risk. The ISO 9000 series uses the term *quality* with a broader meaning, encompassing, besides pharmaceutical quality,

- The interrelationship between the supplier and the customer(s)

- Environmental protection

- Quality assurance in research and development

- Social quality (e.g., conditions at the workplace and leadership practice)

- The necessity to define important company goals (e.g., company image and profitability).

The ISO 9000 series thus overlaps with and is frequently equivalent to GMP regulations. ISO 9000 surpasses GMP, however, when it covers marketing, corrective measures, and the responsibility of management.

The EU adopted the ISO 9000 series as European norms EN ISO 9000 to 9004 in order to establish harmonized systems for product certification and the registration of quality systems. Registration consists of both an audit and the subsequent approval of a quality system against the chosen ISO norm (ISO 9001, 9002, or 9003) by an independent organization, the so-called third-party registrar. For these testing bodies, standards have been developed. Examples include the EN 45000 standards, which govern testing, certification, and the accreditation of notified bodies (32).

ISO 9000 deals not merely with product quality but with quality systems that will lead to sufficient quality (not necessarily to the highest possible quality); it contains information on how a product (or service) is arrived at and not on what is produced.

The series contains recommendations on how to select the relevant norms (ISO 9000) and requirements for the quality system itself (ISO 9001 to 9004). National norms compatible with ISO 9000 exist in most countries (e.g., DIN ISO 9001 to 9003 in Germany and ANSI/ASQC 091 to 093 in the United States) (33).

Certification according to ISO 9000 may be driven by some of the following needs:

- Meeting customers' requirements

- Liability defense

- Quality improvement

- Reduction in rejections by internal quality control

- Higher motivation of employees with regard to quality and quality improvement

Many companies/organizations have installed quality systems and obtained certification. In September 1996, the first pharmacy obtained an ISO certification in Germany (34). There is definitely a strong trend toward this kind of total quality management. However, the advantages of a quality system may also be achieved without necessarily aiming at ISO 9000 certification, as will be set out in the following chapter.

REFERENCES

1. Mathieu, M. 1987. *New Drug Development: A Regulatory Overview*, newly revised and updated edition. Cambridge: Parexel.

2. Hyman, P.M. and Gibbs, J.N. 1996. Top 10 Changes in Regulatory Affairs. *Regulatory Affairs Focus* 1 (9):6–7.

3. Based on Hielscher, M. 1987. Grundzuege des Arzneimittelrechts in der Bundesrepublik Deutschland. In: *Zulassung und Nachzulassung von Arzneimitteln*, edited by B. Schnieders and R. Mecklenburg. Basel: Aesopus Verlag GmbH; pp. 3–5.

4. Gesetz ueber den Verkehr mit Arzneimitteln of 16 May 1961 (AMG 1961, BGBl. I, p. 533).

5. 65/65/EEC (OJ 022 09.02.65, p. 369) (OJ = *Official Journal of the European Communities*).

6. Gesetz zur Neuordnung des Arzneimittelrechts of 24 August 1976 (BGBl. I, p. 2445).

7. Gesetz zur Aenderung des AMG of 24 February 1983 (BGBl. I, p. 169).

8. Gesetz zur Aenderung des AMG of 16 August 1986 (BGBl. I, p. 1296).

9. 83/570/EEC (OJ L 332 28.11.83, p. 1).

10. Walker, S.R. and Griffin, J.P. 1989. *International Medicines Regulation*. Lancaster: Kluwer Academic Publishers.

11. Article 2, Treaty of Rome (signed 25.03.1957 by Belgium, Germany, France, Italy, Luxembourg, and The Netherlands).

12. 65/65/EEC (OJ 022 09.02.65, p. 369).

13. 75/319/EEC (OJ L 147 09.06.75, p. 13).

14. 83/570/EEC (OJ L 332 28.11.83, p. 1).

15. 87/22/EEC (OJ L 015 17.01.87, p. 38).

16. Council Regulation (EC) 2309/93 of 22 July 1993, laying down Community procedures for the authorization and supervision of medicinal products for human and veterinary use and establishing a European Agency for the Evaluation of Medicinal Products (OJ L 214 24.08.93, p. 1).

Council Directive 93/039/EEC of 14 June 1993 (OJ L 214 24.05.93, p. 22), amending Directives 65/65/EEC, 75/318/EEC, and 75/319/EEC in respect of medicinal products.

Council Directive 93/40/EEC of 14 June 1993 (OJ L 214 24.08.93, p. 31), amending Directives 81/851/EEC and 81/852/EEC on the approximation of the laws of Member States relating to veterinary medicinal products.

Council Directive 93/41/EEC of 14 June 1993 (OJ L 214 24.08.93, p. 40), repealing Directive 87/22/EEC on the approximation of national measures relating to the placing on the market of high technology medicinal products, particularly those derived from biotechnology.

17. III/118/87-EN final, January 1989: Notice to applicants for marketing authorizations for medicinal products for human use in the Member States of the European Community.

18. Yakugyo Jiho Co., Ltd. 1994. *Drug Approval and Licensing Procedures in Japan 1994.* Tokyo/Japan; pp. 1–3.

19. D'Arcy, P.F. and Harron, D.W.G. 1992. Proceedings of the First International Conference on Harmonisation Brussels 1991, edited by N. Antrim. Ireland: Greystone Books Ltd.

20. D'Arcy, P.F. and Harron, D.W.G. 1994. Proceedings of the Second International Conference on Harmonisation Orlando 1993, edited by N. Antrim. Ireland: Greystone Books Ltd.

21. Carter, D. 1996. Regulatory Communications: David Carter Reports on ICH3 Discussions. *Regulatory Affairs Journal* 7 (3):184–187.

De Cremiers, F. 1996. Efficacy Workshop, ICH3: Francoise de Cremiers Reviews the Results of the Yokohama Conference. *Regulatory Affairs Journal* 7 (5):359–364.

Harman, R.J. 1996. Medical Regulatory Terminology: Robin J. Harman Outlines the Developments and Latest Proposals Announced at ICH3. *Regulatory Affairs Journal* 7 (2):95–99.

22. Based on Dumitriu, H. 1996. Good Regulatory Practice. *Regulatory Affairs Journal* 7 (10):827–831.

23. Based on a communication received from A.R. Fuell, BIRA, UK of 10 October 1996.

24. Ibid.

25. Leonard, J.D. 1996. Following the Dream (Editorial). *Regulatory Affairs Focus* 9:4.

26. Van Egten, V. 1996. RAPS Europe Helps Breach Barriers (European Update). *Regulatory Affairs Focus* 9:8–9.

27. Parker, J. 1996. Mastering Your Future in Regulatory Affairs. *Regulatory Affairs Focus* 9:21.

28. Heir, R.S. 1994. Good Manufacturing Practice: An Historical Overview and Actual Status. *Drug Information Journal* 28:957–963.

29. Woodward, K.N. 1996. GLP and Veterinary Medicines. *Regulatory Affairs Journal* 8:637–639.

30. Hvidberg, E.F. 1994. An Historical Overview and Actual Status of Good Clinical Practice. *Drug Information Journal* 28:1089–1092.

31. Note for Guidance on Good Clinical Practice (CPMP/ICH/135/95), approved by CPMP on 17 July 1996. Proposed date for coming into operation: Studies commencing after 17 January 1997.

32. Peach, R.W., ed. 1994. *The ISO 9000 Handbook*, 2nd ed. Fairfax, VA: CEEM Information Services.

33. Jackson, P. and Ashton, D. 1995. *ISO 9000—Der Weg zur Zertifizierung*, 2nd ed. Landsberg/Lech: Verlag Moderne Industrie.

34. Erstes ISO—Zertifikat fuer Apotheke. 1996. *Deutsche Apotheker Zeitung* 12 (39):1.

2

Good Regulatory Practice: The Application of Quality Management to Regulatory Affairs[1]

GOOD REGULATORY PRACTICE: A PROPOSAL FOR A QUALITY SYSTEM

The problem today seems to be the industry's insufficient quality of submissions to Regulatory Bodies. And this problem apparently cannot be solved by trying to add quality to the submissions through a flood of regulations, guidelines, position papers, and internal Standard Operating Procedures (SOPs) that Regulatory Bodies and the industry alike are producing. Efforts toward harmonization (as attempted by the International Conference on Harmonisation [ICH]) and calls for deregulation have been unsuccessful in improving this situation. The underlying problem appears to be that the industry and Regulatory Bodies too often perceive each other as opponents rather than as partners with common interests and definitely not as part of a holistic quality system. Considering how many (repetitive) control steps are presently installed by companies (e.g., quality assurance, technical writing, peer review), and Regulatory Bodies (e.g., dossier check-in controls), the present system, based on mutual distrust,

---

1. Parts of this chapter were based on Dumitriu, H. 1996. Good Regulatory Practice. *Regulatory Affairs Journal* 7 (10):827–831.

obviously involves high costs in terms of time, manpower, and rejects on both sides without real quality improvement. Further indicators of system malfunction are as follows:

- Revealing all rather than crucial deficiencies

- Review of varying quality and/or time

- Additional control steps or even complete control departments being established

Good Drug Regulatory Practice (GDRP) might be the solution to this problem.

The term *Good Regulatory Practice* (GRP), although nonofficial, has been with us for some time, first used in a clinical context (1). Regulatory Affairs professional societies (e.g., the Central European Regulatory Affairs Society (MEGRA) and the Pan-European Federation of Regulatory Affairs Societies (PEFRAS), have been working on the subject of GRP guidelines to define more clearly the role of the Regulatory Affairs manager (2). Regulators have used the term *GRP* when calling for guidelines by Regulatory Bodies in order to give applicants assurance about the quality of the review and of further procedures within the Regulatory Bodies and describing quality expected from the applicant (3).

In this book, GDRP in its broadest sense means the definition and establishment of a quality system that comprises both the industry and the Regulatory Bodies. In fact, it is about turning the focus away from trying to achieve quality by more control to producing quality in the first place. As companies strive to improve quality, quality management on the part of the Regulatory Bodies should be as important as the industry's efforts.

What is a quality system? It is deciding what the goals are, and what must be done by whom and how in order to reach these goals, writing this down, adhering to it, watching the results, and modifying the system, if required, for improvement. Specifications for materials/services received from other parties and specifications for products/work results/services should be discussed and agreed on together by both the supplier(s) and the customer(s). A quality system can clarify the requirements, agree on contracts, build up confidence in the quality produced by the other party, and eliminate, to a certain extent, the duplication of tests.

What are the ultimate goals of GDRP for Regulatory Bodies and the industry? Obviously, quick approvals for medicinal products with proven quality, safety, and efficacy. It then follows that quick rejections of medicinal products not fulfilling these criteria are needed. As the resources of Regulatory Bodies are not unlimited (whose are?), industry (and taxpayers) must be certain that the available resources are spent in an efficient way. This means that Regulatory Bodies must make sure that they do not waste time on unapprovable medicinal products. In addition, the industry wants Regulatory Bodies to achieve and preserve an image of high standing. This point may seem arguable, as high standing tends to equal tough requirements and bigger efforts to meet those requirements. However, considering that the image of a particular Regulatory Body directly impacts the willingness of other Regulatory Bodies to accept/recognize its decisions, assessment reports, input as rapporteur, inspections, free sale certificates, dossier formats and so on, this is also a worthwhile issue for the industry. A good image of a Regulatory Body is a major advantage for any national industry and helps increase the return on investment as it increases the value of company portfolios by casting favorable and universally accepted decisions.

What are the goals of GDRP?

- Efficiency: quick and qualified decisions on the quality, safety, and efficacy of products

- Effectiveness/productivity: effective use of resources, cost-effectiveness

- Results: achieve and preserve an image of high standing

What are the advantages of GDRP? The joint goals of making quality medicinal products available to patients in a timely fashion will be reached by a holistic quality system rather than by the present way of doing things, because the global picture is seen and the entire process is designed to produce quality.

Goals will be perceived as being of mutual interest for all interested parties and, therefore, efforts can be combined. The quality system will provide for continuous improvement. It will also lead to a reduction of duplicate testing, thereby saving time, manpower, and resources. One might consider this suggestion against the concept of equal treatment for all applicants (and Regulatory Bodies). However, this argument is not valid, as the same high quality will be expected from all applicants. Only those companies that provide detailed insight into their procedures as well as proof that their quality system ensures adequate quality on a continuous basis will benefit, as less confirmation will be required at a later stage of the marketing authorization process. The same applies to Regulatory Bodies: The quality system should lead to a high degree of recognition of their decisions by other Regulatory Bodies and, as a result, to a decrease in the need to review again or conduct additional inspections. Eventually, the quality system will lead to the right regulation, whereas conflicting or outdated regulations will be abolished.

What are the advantages of GDRP for Regulatory Bodies?

- Reaching the goals

- Looking at the global picture and seeing the whole process

- Common interest(s) of Regulatory Bodies and industry

- Continuous quality improvement

- Reduction of duplicate tests for certified companies and Regulatory Bodies maintaining a Quality System (saving time/manpower/costs)

- Abolish conflicting or outdated regulation/right-sizing regulations

GOOD REGULATORY PRACTICE: A PROPOSAL FOR A QUALITY SYSTEM FOR REGULATORY BODIES

The following is a proposal for a generic quality system with the aim of starting a discussion on the subject within the industry and Regulatory Bodies. The quality system should cover the following topics:

- Archiving

- Assessment reports

- Auditing and compliance

- Certification

- Authorization process/Change alert

- Clinical trial license
- Contact reports
- Contact with industry
- Crisis management
- Electronic data processing (EDP)/Telecommunication
- Education/Training
- EU procedures
- Environmental protection
- Import/Export (if applicable)
- Information management
- Inspection
- Joint review
- Marketing authorization application
- Periodic safety report
- Pharmacovigilance
- Policy
- Pricing (if applicable)
- Product labeling
- Project assignments
- Promotion/Advertising compliance (if applicable)
- Regulations and guidelines
- Renewal of marketing authorization
- Review
- Standardization (format, content requirements)
- Terminology

Additional topics may be added during the discussions, as some participants may have additional or different preferences.

These topics should be documented in a Quality Manual available at each employee's workplace. The Quality Manual should also be agreed to by industry. It would be desirable to agree at the World Health Organization (WHO)/ICH/International

Federation of Pharmaceutical Manufacturers Associations (IFPMA) level on an internationally accepted part (global policies, standards), and then to develop a second, national, or local part for each Regulatory Body that translates the global part into local SOPs tailored to the specific needs of that Regulatory Body. Keeping this in mind, the table of contents of a Quality Manual for Regulatory Bodies might look as follows:

1   INTRODUCTION

   1.1 Instructions for use

   1.2 Distribution

   1.3 Purpose and scope

   1.4 Regulatory Authority structure and organization

   1.5 Description of functional units

2   QUALITY POLICY

   2.1 General statement

   2.2 Policy statement

   2.3 Details of Regulatory Body structure and organization

   2.4 Statement of authority and responsibility

3   THE QUALITY SYSTEM'S OUTLINE

4   INDEX

   4.1 Policies

   4.2 Standards

   4.3 Standard Operating Procedures

5   POLICIES

6   STANDARDS

7   STANDARD OPERATING PROCEDURES

However, the procedures of Regulatory Bodies are beyond the scope of this book. Instead, the procedures of the industry and especially of Regulatory Affairs professionals/departments work will be emphasized.

GOOD REGULATORY PRACTICE: THE ROLE OF REGULATORY AFFAIRS IN PRODUCT DEVELOPMENT

This section focusses on the value contributed by high quality regulatory affairs work to product development and discusses some obstacles that the Regulatory Affairs professional in the industry may be confronted with, especially when trying to establish a quality system.

The functions of Regulatory Affairs personnel are to act as follows:

- The outlet of the company toward Regulatory Bodies

- Interpreter of regulations to the company

- Influencer of new regulations (i.e., as THE interface between the company and Regulatory Bodies)

However, Regulatory Affairs was not always seen like this. Why? The most common prejudices that Regulatory Affairs departments are confronted with are as follows:

- We have always done it like this.

- Anybody with common sense can do it.

- We are the experts.

History

As the regulatory environment has evolved, more and more people within companies have become exposed to regulatory requirements. This sometimes makes them believe that clinical trial licenses or marketing authorizations are easier to obtain than the Regulatory Affairs department indicates. Unfortunately, the Regulatory Affairs staff is almost always right, as they know that they are trying to hit a moving target. Any experience, though valuable in itself, will probably be outdated after one to two years at the most. Therefore, whatever experience has been accumulated, it must be checked continuously against new regulations. Hence, the growing importance of Regulatory Intelligence and Information Services.

Education

Academic or university degrees are considered as proof of the capability to do highly qualified work. Regulatory Affairs, however, is not generally taught at universities (though there seems to be a trend in the United States toward offering postgraduate qualification and training courses in regulatory affairs). The work is thus erroneously considered easy—a job that can be done by almost anyone and almost without training.

This aspect has been taken up by Regulatory Affairs professional societies (e.g., the British Institute of Regulatory Affairs [BIRA], the Regulatory Affairs Professional Society [RAPS], etc.) who offer educational or certification programs. However, regulatory affairs is not a subject that can be taught in a purely theoretical way; it also deals with experience gained on the job. This learning by doing must include contact with Regulatory Bodies, not so much to gain knowledge but to develop a certain instinct, which some succeed in developing and others, unfortunately, will never achieve. It is this instinct that really makes the difference in an environment, where common sense will simply not do the job. Thus, it will take at least three years of continuous involvement to become a professional in the field of Regulatory Affairs.

Scientific and Technical Experts

Scientific and technical experts are probably the most difficult group of prejudiced people with whom the Regulatory Affairs professional is confronted. There are always experts who know better. The solution of the conflicts between experts and Regulatory Affairs professionals is simple. Experts are needed, but more in the way of last resort. Let Regulatory Affairs first check the issue in the context of the entire submission. The Regulatory Affairs professional can act as an "all-around-manager", overseeing all aspects of the dossier and of all single documents, as well as acting as a scout through more than 100,000 pages and more of documents. Call in the experts only if there really is a need for a scientific discussion. While the needs of the reviewer really may have been quite simple (e.g., to locate some special information), the issue may become more complicated or problems that were not even addressed by a reviewer will suddenly surface, as a result of the expert's awareness of it. Also, the company's experts tend to believe that they and experts in the Regulatory Bodies think along the same lines. This may be true in some respects, but this means forgetting that experts from industry speak their company's language simply by adapting to this special environment, just as experts of the Regulatory Body speak their own language. From this results a need for the Regulatory Affairs professional to be an interpreter of guidelines, letters of deficiencies, and so on. Once the importance of the Regulatory Affairs function is recognized, the question becomes: What is GDRP?

Good Regulatory Practice

The nonofficial term *Good Regulatory Practice* was originally used in a clinical context, first by Regulators to signify guidelines ruling the Regulatory Body's work in giving applicants assurance about the quality of the review by and the procedures of Regulatory Bodies (4), and then by Regulatory Affairs professional societies to define more clearly the role of Regulatory Affairs managers. It is this last aspect that will be discussed in this book. A quality system will be proposed for Regulatory Affairs departments as being a vital part of an overall quality system encompassing both the industry and Regulatory Bodies.

What is GRP? It is an overall quality system for handling the regulatory aspects of medicinal products (i.e., Good Practice of Regulatory Bodies and Good Practice of Regulatory Affairs departments in the industry). But what is high quality in Regulatory Affairs, or, in other terms, how can a Regulatory Affairs department's performance be checked?

Excellent communication skills are important, but there must also be a general attitude that considers internal scientific and technical disciplines and Regulators as partners, not as enemies. Regulatory Affairs should work prospectively—take an active part in forming the regulatory environment by participating in Working Parties and by commenting on evolving guidance documents. It is especially important to be well informed about the direction of developing regulations and guidelines.

In project teams, Regulatory Affairs should play a proactive role by suggesting submission strategies, discussing the pros and cons, and informing and updating the project team on the current status of regulations, while preferring pragmatic approaches over 100 percent (or perfect) solutions. Regulatory Affairs should screen regulations and guidelines as well as the competitor's activities in a market-oriented way and point out business opportunities from a regulatory perspective (e.g., by extending claims). The main function, of course, is the output of customer-oriented dossiers meeting (but not exceeding) regulatory requirements in a timely fashion.

Project teams should ask for regulatory expertise and invite comments, because medicinal product development is not only about speeding up the process to submission. It is primarily about knowing in which global direction to head. This requires a paradigm shift by focussing on final approval in all major countries, which mandates the incorporation of the activities during the review cycles into the plans.

Some areas where Regulatory Affairs might be of specific value to companies to avoid current mistakes are listed below to give an idea of the pitfalls of medicinal product development:

- Calling in Regulatory Affairs after finalization of the development/marketing plan.

- Suppressing critics in the organization.

- Hiding critical issues.

- Telling all you know.

- Trying to make it perfect.

- Doing all the studies as early as possible.

- Using in-house methods to structure the documentation.

All of these mistakes will cost the company a lot of time and money, but by far the worst mistake is to call in Regulatory Affairs only after all of the plans for development and marketing have been finalized. If this mistake can be avoided, Regulatory Affairs will do its best to help avoid most of the other pitfalls.

Telling everything leads to confusion. Tell only relevant things. On the other hand, do not try to make it too perfect. A lot of time will be lost and there is a very real danger of raising the standards for future applications. A similar mistake is to do all of the studies, including those for marketing purposes, BEFORE the first submission. Of course, having started them, you must report the results, which may not necessarily support your efficacy claims but instead cause problems on the safety side. Keep in mind that the more patients who receive a medicinal product, the longer the list of possible side effects, interactions, and so on.

Be frank about problems. This is where critics within the organization are very valuable in identifying what these problems are, while everybody else is very enthusiastic about the product and, therefore, is inclined to see only the sunny sides of it. NEVER try to hide anything. Chances are good that a reviewer will spot it. Then, in addition to being confronted with the obvious problem, the company will earn a bad reputation in the eyes of the reviewers and may be subjected to a much closer and longer review in the future. Also, there will be a tendency to suspect problems even where there are none.

Another pitfall may be presenting the documentation in the same format as the company has always done it. Regulators usually accept this, unless the order is too unusual. However, it will cost the company time during dossier check-in and review, and the company may loose some of the goodwill of the reviewer. He or she may even prefer to look at another company's documentation first, and the company will never learn why the review of the application took so long. Therefore, instead of trying to make the Regulators adapt to your structure, use theirs. Use their suggested table of contents, their headings, and their terminology as extensively as possible.

The project team should make use of Regulatory Affairs! The Regulatory Affairs professional is one of the most valuable team members, if allowed to live up to full capacity. If the Regulatory Affairs department is not proactive, challenge them. Ask for project-specific evaluations and scenarios, including regular updates on the status of regulations during medicinal product development. Request Regulatory Affairs service or input with regard to checklists of required documents, document tracking, and check-in controls.

Each project requires the following information:

- Evaluation of the regulatory environment

- Checklist of required documents

- Scenarios for clinical trial license and marketing authorization strategy (document tracking, check-in controls)

- Regular updates on the above

In order to provide high quality Regulatory Affairs work, the mission of Regulatory Affairs in the organization must be clear. A quality system should be in place that defines and assures quality, standardizes products/work results and procedures, and fosters continuous quality improvement.

The following is a checklist for the performance of Regulatory Affairs departments:

- Perceives disciplines and Regulators as partners, not as enemies

- Establishes/maintains efficient contacts with Regulatory Bodies

- Works proactively

- Proactive in medicinal product development/maintenance teams

- Market oriented and customer focussed

- Submits dossiers of sufficiently high quality in a timely fashion to obtain and maintain marketing authorizations

- Maintains a quality system

The following topics should be covered by the quality system: archiving, auditing and compliance, change alert/authorization process, clinical trial license application, contact reports, contact with Regulatory Body, crisis management, documents for regulatory purposes, dossier, EDP/telecommunication, education/ training, (electronic) submission, environmental protection, import/export, information management, inspection, labeling, marketing authorization application, periodic drug safety report, policy, project assignments, promotion/advertising compliance, regulations and guidelines, and terminology. The elements of the quality system should be documented in a Quality Manual. A copy of the Quality Manual should be made available to each Regulatory Affairs employee. International companies are well advised to develop an international part of the Quality Manual containing global policies and standards that are compulsory for everybody in the global Regulatory Affairs organization. Subsequently, each local or national Regulatory Affairs unit should establish a second or local part that translates the global policies and standards into local SOPs tailored to the specific needs of that unit. Part II of this book is a proposal for such a Regulatory Affairs Quality Manual.

REFERENCES

1. Jefferys, D.B. 1992. Good Clinical Practice: EC Expectations, Education, Self-Regulation, or Imposition? *Drug Information Journal* 26 (4):609–613.

   Pasotti, V. 1994. Local and International Regulatory Aspects of Multicentre Trials. *European Journal of Clinical Research* 6:18–25.

2. Tryzelaar, B. 1989. Regulatory Affairs and Biotechnology in Europe. I. Introduction into Good Regulatory Practice. *Biotherapy* 1 (1):59–69.

   Europaeische Gesellschaften finden unter dem Dach von PEFRAS zusammen, 20.04.1995, *Arzneimittel-Zeitung*, 8/7A.

3. Baccouche, M., and Schweim, H. 1995. Good Regulatory Practices. *Regulatory Affairs Journal* 6 (7):547–551.

4. Jefferys, D.B. 1992. Good Clinical Practice. EC Expectations, Education, Self-Regulation, or Imposition? *Drug Information Journal* 26 (4):609–613.

   Pasotti, V. 1994. Local and International Regulatory Aspects of Multicentre Trials. *European Journal of Clinical Research* 6:18–25.

3

Check Your Quality System

In order to allow you to quickly evaluate the possible benefit of establishing a quality system as set out in Part II and to bench mark your present system, the following questionnaire is provided. It has been developed to enable you to check your present quality system easily and quickly and to identify potential deficiencies. The questionnaire can also be used to check whether individual Regulatory Affairs employees have sufficient knowledge and understanding of an already existing quality system. Furthermore, it may be used repeatedly at regular intervals in order to control and maintain a high degree of quality and understanding in the quality system. For conclusions from the test results, turn to page 44.

QUALITY SYSTEM QUESTIONNAIRE

Name: _____   Department: _____   Date: _____

Fill in the questionnaire, ticking "yes", "no", or "not sure" as appropriate. Do not take too long to do this—answers should be spontaneous.

Question	Response			Policies That Address This Issue
Are the project team's targets defined with regard to target summary of product characteristics (SMPC), target countries, applicant(s), license holder(s), trademark(s), dosage form(s), strength(s), primary packaging(s)?	yes ☐	no ☐	not sure ☐	26
Are Regulatory Affairs employees motivated with regard to environmental protection?	yes ☐	no ☐	not sure ☐	12, 14
Are you involved in industry Working Parties that actively participate in discussion processes concerning rule making by Regulatory Bodies?	yes ☐	no ☐	not sure ☐	25
Are you sure that the labeling of your medicinal products in all markets reflects your company's actual state of knowledge?	yes ☐	no ☐	not sure ☐	20
Are you sure that during the life cycle of your company's medicinal products you have a good overview of all product changes, including labeling changes, in order to keep the documentation harmonized?	yes ☐	no ☐	not sure ☐	15
Are there processes in place to guard against electronic data processing (EDP) security problems?	yes ☐	no ☐	not sure ☐	18
Are you using computers for Regulatory Affairs work other than word processing (e.g., electronic databases for document management, registration status, document tracking, patient leaflet information)?	yes ☐	no ☐	not sure ☐	29

Table continued on next page.

Table continued from previous page.

Question	Response			Policies That Address This Issue
	yes ☐	no ☐	not sure ☐	
Are procedures in place to ensure that you are in compliance with your policies/standards/SOPs?	yes ☐	no ☐	not sure ☐	2
Are your company's medicinal products in compliance with registered information (e.g., labeling, composition, manufacturer)?	yes ☐	no ☐	not sure ☐	6
Are you sure you always understand what the Regulators tell you during contact with their Regulatory Body?	yes ☐	no ☐	not sure ☐	7, 8
Are procedures agreed on for communicating information on contacts with a Regulatory Body within the company?	yes ☐	no ☐	not sure ☐	7
Are responsibilities within Regulatory Affairs clear to all employees and also to disciplines, departments, affiliates, and licensees with whom you work?	yes ☐	no ☐	not sure ☐	23
Are contacts with a Regulatory Body coordinated by a control desk or a responsible person/department?	yes ☐	no ☐	not sure ☐	8
Are processes in place to protect your data (including personal data)?	yes ☐	no ☐	not sure ☐	18
Are you sure that nobody is tampering with personal data captured by EDP?	yes ☐	no ☐	not sure ☐	18
Are you sure that actually used patient leaflets concerning your company's medicinal products (also in foreign languages) inform patients adequately?	yes ☐	no ☐	not sure ☐	20
Are your databases protected from viruses?	yes ☐	no ☐	not sure ☐	18
Are key terms used with the same meaning within Regulatory Affairs/the company?	yes ☐	no ☐	not sure ☐	28
Are there established standards for the generation of dossiers in your department?	yes ☐	no ☐	not sure ☐	11

Table continued on next page.

Table continued from previous page.

Question	Response			Policies That Address This Issue
Can Regulatory Affairs submit applications in every country within less than one month from receipt of the dossier/documentation?	yes ☐	no ☐	not sure ☐	27
Can you swear that all promotion/advertising for your company's medicinal products is in compliance with registered labeling and local legal requirements?	yes ☐	no ☐	not sure ☐	24
Can you provide on-the-spot copies of all presently used package leaflets of your company's medicinal products?	yes ☐	no ☐	not sure ☐	5
Can you produce the presently valid documentation for all of your company's medicinal products on the spot (e.g., for an inspection)?	yes ☐	no ☐	not sure ☐	15
Could you submit in less than one month renewal applications, including documentation, for all of your company's medicinal products in one country?	yes ☐	no ☐	not sure ☐	15
Do you know what to do if you find that an important product change requiring prior approval has already been carried out and unauthorized product is being sold?	yes ☐	no ☐	not sure ☐	9
Do you know what to do if you have missed the deadline for the application of the renewal of a marketing authorization for your company's biggest product in a major market?	yes ☐	no ☐	not sure ☐	9
Do you have all of your Regulatory Affairs policies/standards/SOPs available in case of an inspection?	yes ☐	no ☐	not sure ☐	2
Do you know how many cubic metres of archived documents you have?	yes ☐	no ☐	not sure ☐	5
Do the disciplines provide you with documents for regulatory purposes that have gone through a quality assurance process prior to being shipped to Regulatory Affairs?	yes ☐	no ☐	not sure ☐	10

Table continued on next page.

Table continued from previous page.

Question	Response			Policies That Address This Issue
Do you have a good knowledge of the Contract Research Organizations (CROs) available for Regulatory Affairs work, and the quality of their work, cost effectiveness, and reliability?	yes ☐	no ☐	not sure ☐	21
Do you comment on draft guidelines to industry association(s)/Regulatory Body(ies)?	yes ☐	no ☐	not sure ☐	25
Do you frequently receive phone calls from people hunting for the responsible Regulatory Affairs manager for a specific medicinal product?	yes ☐	no ☐	not sure ☐	23
Do you closely monitor the actual regulatory environment in all relevant countries and on a global level (ICH, WHO, major EU, or U.S. developments)?	yes ☐	no ☐	not sure ☐	25
Does your company save money through environmental protection measures in your department?	yes ☐	no ☐	not sure ☐	14
Does everybody within Regulatory Affairs/the company understand the abbreviations for key Regulatory Affairs terms?	yes ☐	no ☐	not sure ☐	28
Does Regulatory Affairs provide different scenarios to the project team (e.g., with regard to wording of claims, procedures to be used)?	yes ☐	no ☐	not sure ☐	26
Does every Regulatory Affairs employee have the quality policies, standards, SOPs available at his or her workplace and does he or she use them in everyday work?	yes ☐	no ☐	not sure ☐	12
Has responsibility for contact with the Regulatory Body been defined?	yes ☐	no ☐	not sure ☐	8
If you are using software programs or databases in Regulatory Affairs, is there an added value to the Regulatory Affairs department or company?	yes ☐	no ☐	not sure ☐	29
If the department head had a heart attack, would you know what to do? Could you replace him or her?	yes ☐	no ☐	not sure ☐	9, 12

Table continued on next page.

Table continued from previous page.

Question	Response			Policies That Address This Issue
Imagine your whole building burned down today. Would you again be operative within the next two working days?	yes ☐	no ☐	not sure ☐	5, 9
Imagine that key personnel in submission management became ill. Would Regulatory Affairs still be able to do the same high quality submissions?	yes ☐	no ☐	not sure ☐	27
In case of an EDP security problem, would it be possible to identify the culprit?	yes ☐	no ☐	not sure ☐	18
In case of a crisis, would you know which CRO to turn to for help?	yes ☐	no ☐	not sure ☐	21
In case of absence (e.g., illness, vacation), do you give access to your E-mail or databases to authorized person(s) (e.g., department head, secretary) rather than passing on your password(s)?	yes ☐	no ☐	not sure ☐	18
Is there a uniform format for Regulatory Affairs policies/standards/SOPs?	yes ☐	no ☐	not sure ☐	1
Is it generally true that an electronic submission could delay an application for a marketing authorization for as much as half a year?	yes ☐	no ☐	not sure ☐	13
Is Regulatory Affairs work governed by policies, standards, or SOPs?	yes ☐	no ☐	not sure ☐	1
Is layout regarding format, structure, and content of submissions in a country similar (or does each Regulatory Affairs employee create his or her own style)?	yes ☐	no ☐	not sure ☐	27
Is promotion/advertising compliance the responsibility of Regulatory Affairs in your company?	yes ☐	no ☐	not sure ☐	24
Is it generally true that an electronic submission might put an application for a marketing authorization into the top priority group for review?	yes ☐	no ☐	not sure ☐	13
Is the education/training of Regulatory Affairs employees monitored continuously?	yes ☐	no ☐	not sure ☐	12

Table continued on next page.

Table continued from previous page.

Question	Response			Policies That Address This Issue
	yes	no	not sure	
Marketing authorization application(s) of your company have never been rejected.	☐	☐	☐	4
Marketing authorization application(s) of your company have never been delayed because of a failure/delay in obtaining clinical trial license(s).	☐	☐	☐	3
Most/all documents for regulatory purposes are written by the scientific disciplines (not Regulatory Affairs).	☐	☐	☐	10
Once a product is marketed, do you have the periodic Safety Update Report available?	☐	☐	☐	22
Projects have never been delayed by the absence (e.g., illness, vacation) of the responsible Regulatory Affairs manager.	☐	☐	☐	17
Regulatory Affairs in your company renounces double-checking each document for regulatory purposes (e.g., format, structure and content, typing errors).	☐	☐	☐	10
Studies have never been delayed due to problems with the import of study medication.	☐	☐	☐	16
There has never been a shortage of product on the market due to export problems from the country of manufacture.	☐	☐	☐	16
There has never been an unauthorized disclosure of confidential information or distribution of wrong information.	☐	☐	☐	17
When an inspection is announced, this does not cause special excitement in the company.	☐	☐	☐	19
Would you feel embarrassed if suddenly confronted with your company's patient leaflets used in developing countries?	☐	☐	☐	20
You have never lost/misplaced an original marketing authorization document.	☐	☐	☐	5

CONCLUSIONS FROM THIS TEST

In all cases where you have answered "no", you definitely have a problem. If you have answered "not sure", this may indicate a problem or at least a lack in the transparency of your quality system. If you wish to work on the most pressing deficiencies of your quality system, turn directly to the policy or policies indicated for the respective question in the questionnaire. However, make sure that you eventually work through the entire Quality Manual.

Part II

THE REGULATORY AFFAIRS QUALITY MANUAL

4

The Quality Manual Explained

INTRODUCTION

Instructions for Use

The Quality Manual should be read carefully in order to obtain a complete picture on the purpose and the scope of the quality system. For special topics, see the policies. Revisions should be integrated immediately after receipt in order to keep the Quality Manual up-to-date.

Distribution

The Quality Manual is distributed by Regulatory Affairs within Regulatory Affairs to all employees. The Quality Manual should be available to each Regulatory Affairs employee at his or her workplace. Distribution and update lists will be maintained showing who holds which copy number of the Quality Manual and the issue status of each copy. If possible, the Quality Manual may also be made available electronically to all Regulatory Affairs employees. However, such electronic systems must be validated in order to guard against unintentional modifications of the texts.

Any Regulatory Affairs employee can suggest modifications or improvements. Decision making will be at the highest decision level by the Quality Steering Committee. Regulatory Affairs will produce and distribute revisions to the Quality Manual in a timely fashion. Revisions will be accompanied by a cover sheet identifying pages to be added, replaced, or deleted.

Each recipient will be required to return the signed and dated cover sheet with the statement that the changes have been integrated. This information will be reflected in the distribution and update lists.

Purpose and Scope

The purpose of this Quality Manual is to define the quality of work results, such as dossiers, services, and procedures for Regulatory Affairs worldwide.

Regulatory Affairs Structure and Organization

The functions of Regulatory Affairs (internationally operating companies, adjust to your organization) are as follows:

- Quality Steering Committee
- Corporate Regulatory Affairs
- Regulatory Intelligence
- European liaison
- National Regulatory Affairs

Responsibility of Functional Units

1. Quality Steering Committee: decision making on quality management and quality system

2. Corporate Regulatory Affairs:

   - Accompanying the medicinal product development process and providing regulatory input (e.g., generating regulatory strategies)

   - Providing dossiers (1), and, if applicable, global dossiers (2), to national Regulatory Affairs

3. Regulatory Intelligence: monitoring the regulatory environment, interpreting guidelines, influencing evolving regulations

4. European liaison: interface, facilitator, coordinator regarding the EU procedures, and contact with the European Agency for the Evaluation of Medicinal Products (EMEA)

5. National Regulatory Affairs: generating submissions (3) based on dossiers, or global dossiers; managing the submission process and maintenance for the company's medicinal products

REGULATORY AFFAIRS QUALITY POLICY

General Statement: Company Guidelines for Quality Management

(Insert your company's quality statement. The following subjects should be covered: quality, employees, quality as a top management function, quality management Steering Committee.)

Regulatory Affairs Quality Policy Statement (4)

Quality means fitness for the intended purpose in all aspects of regulatory affairs. Regulatory Affairs will strive to meet the needs of its internal and external customers efficiently through a continuous quality improvement process and development and maintenance of a quality system.

- All Regulatory Affairs employees are requested to comply with the provisions of the quality system as documented in the Quality Manual.

- All Regulatory Affairs employees are responsible for quality improvement and development and maintenance of the quality system.

- All Regulatory Affairs employees are invited and allowed to participate in quality improvement activities.

- There will be a measurable annual assessment of the quality of Regulatory Affairs, and, if required, a quality improvement process will establish defined quality objectives for the quality system.

- The quality system will be implemented systematically in every part of Regulatory Affairs.

- Education and training are vital to the quality improvement process and the development and maintenance of the quality system.

- Emphasis must be on prevention rather than on control, quality built in, not inspected in.

- Regulatory Affairs will involve internal as well as external suppliers and customers as far as possible in the development and maintenance of the quality system and the quality improvement process.

- A Quality Steering Committee will be responsible for decision making on quality management and the quality system and coordination of the quality improvement process.

Authorization: _____

Effective date: _____

Implementation date: _____

Details of Regulatory Affairs Structure and Organization

Quality Steering Committee

> Responsibility: Decision making on quality management and the quality system

Because quality management is the top management task, this committee should consist of the highest authorized persons required to ensure empowerment for developing and maintaining the quality system as well as for the continuous quality improvement process. The head(s) of Regulatory Affairs might elect quality controllers to be supporting members of the Quality Steering Committee.

Corporate Regulatory Affairs

> Responsibility: Accompanying the medicinal product development process and providing regulatory input (e.g., generating regulatory strategies), providing dossiers or global dossiers to national Regulatory Affairs

Corporate Regulatory Affairs will typically be located at the product development site of the company. The function will be performed by very experienced Regulatory Affairs professionals. Ideally, they should have previously had a function in national Regulatory Affairs of one or two major countries and also have experience with the EU procedures (i.e., having acted in the European liaison function) in order to enable them to be efficient project team members

Regulatory Intelligence

Responsibility: Monitoring the regulatory environment, interpreting guidelines, influencing evolving regulations

Typically, Regulatory Intelligence is a task for a small group of specialists, knowledgeable in the sources of regulatory information relevant to the specific company and the development of its medicinal products. Monitoring the regulatory environment includes screening literature, also via electronic libraries and the Internet; maintaining a database or repository for relevant guidelines; distributing information within the organization; and evaluating and interpreting guidelines to the company. On a higher level, Regulatory Intelligence will also mean influencing evolving regulations by actively participating in Working Parties and commenting on draft guidance documents (in collaboration with the scientific disciplines concerned).

European Liaison

Responsibility: Interface, facilitator, coordinator regarding the EU procedures, and contact with the EMEA

The European liaison function is usually performed by a very small group of very experienced Regulatory Affairs professionals (ideally, they should have previous national regulatory experience from one or two EU Member States) with some technical assistance (the Secretariat). They are typically seated close to the EMEA, as their job is to develop and maintain excellent contacts with this agency and to steer the EU procedures for the company. With regard to applications in the EU, they act as the interface between corporate Regulatory Affairs and national Regulatory Affairs departments.

National Regulatory Affairs

Responsibility: Generating submissions based on dossiers or global dossiers; managing the submission process and maintenance for the company's medicinal products

Regulatory Affairs staff performing this function should develop and maintain excellent contacts with the national Regulatory Body and be knowledgeable about local regulatory requirements. They should have excellent command of the national language.

Statement of Authority and Responsibility

The responsibility, authority, and interrelation of all Regulatory Affairs personnel who manage, perform, and verify work affecting quality is defined by job descriptions that are to be updated as required. The actual job descriptions are available at the pertinent personnel department.

THE QUALITY SYSTEM'S OUTLINE

The quality system covers the quality of work results, the quality of processes, the social quality, and the environmental quality. It consists of policies, standards, and Standard Operating Procedures (SOPs). Policies define the basic principles under which Regulatory Affairs is to operate. Standards are definitions of items that are required to be identical throughout the Regulatory Affairs organization. SOPs define how policies are implemented and standards are met in daily operations.

Decision making on policies, standards, and SOPs is by the Quality Steering Committee. Authorized person(s) are responsible for the implementation of the policies, standards, and SOPs and for appropriate compliance with their provisions. This will be adequately documented by developing the corresponding SOPs.

Quality will be defined for each work result (e.g., dossier) and/or service or process (e.g., auditing) by the policies. It will be defined in a measurable way (as far as possible) by standards and/or SOPs. For each work result, as far as possible, together with internal as well as external supplier(s) and customer(s), the necessary quality of input and output will be defined, focussing on

- Specifications (e.g., critical items to be controlled)
- Limits or ranges of tolerance (e.g., in a dossier, number of missing pages)
- Frequency/extent of checks
- Documentation of check results
- Responsibility for such checks
- Necessity for duplication of checks by customer(s)

Quality training will be continuously available for all Regulatory Affairs employees. Continuous feedback is an important part of the quality system. Responsibility for feedback lies with each individual employee.

Once per year, an audit of all Regulatory Affairs policies, standards, and SOPs will be performed. It will be preannounced. An audit plan will be written and distributed in advance. The results of the audit will be documented in an audit report. Based on the principle of objectivity, there should be representations of supplier(s) and customer(s) during the audit. Additionally, a neutral third party should be present. If deficiencies are identified, adequate measures will be taken to improve the quality of the particular work result or process.

The Quality Steering Committee will report to upper management on the status of the quality system once per year. The report will be signed by authorized person(s).

NOTES

1. Dossier: In this book, "dossier" will be used to signify a compilation of documents relevant to a specific regulatory purpose (e.g., application for a clinical trial authorization or application for a marketing authorization) in a specified country(ies) for a developmental or marketed medicinal product in a structured form. It is a subset of the global dossier. The dossier is the basis for the submission.

2. Global dossier: In this book, "global dossier" will be used to signify a compilation of all documents required for international regulatory purposes for a developmental or marketed medicinal product.

It is maintained continuously throughout the life cycle of the medicinal product and serves as a repository for the generation of dossiers and submissions.

3. Submission: In this book, "submission" will be used to signify a country-specific compilation of documents for a specific regulatory purpose (e.g., application for a clinical trial license or application for a marketing authorization) for a developmental or marketed medicinal product in a structured form according to national regulatory requirements. It is based on the dossier or, if applicable, the global dossier. It may contain additional national documents (e.g., national leaflets or application forms).

4. Adapted from Peach, R.W., ed. 1994. *The ISO 9000 Handbook*, 2nd ed., Fairfax, VA: CEEM Information Services.

APPENDIX: NUMBERS AND TITLES OF POLICIES AND STANDARDS IN THIS MANUAL

Policies

01: Policy

02: Auditing and Compliance

03: Application for Clinical Trial License

04: Application for Marketing Authorization

05: Archiving Management

06: Change Alert/Authorization Process

07: Contact Report

08: Contact with Regulatory Body

09: Crisis Management

10: Documents for Regulatory Purposes

11: Dossier

12: Education/Training

13: Electronic Submission

14: Environmental Protection

15: Global Dossier

16: Import/Export

17: Information Management

18: Information Technology

19: Inspection

20: Labeling

21: Outsourcing

22: Periodic Safety Update Report

23: Project Assignments

24: Promotion/Advertising Compliance

25: Regulations and Guidelines

26: Regulatory Strategy

27: Submission

28: Terminology

29: Tools

Standards

01.01:   Policy

01.02:   Standard Operating Procedure

03.01:   U.S. Application for Clinical Trial License:
IND Content and Format

04.01:   EU Application for Marketing Authorization:
Chemical Active Subsatnce(s)

04.02:   EU Application for Marketing Authorization:
Biological(s), Part II

04.03:   U.S. Application for Marketing Authorization:
NDA Content and Format

07.01:   Regulatory Body Contact Report

10.01:   Regulatory Document Types

11.01:   Dossier

15.01:   Global Dossier

20.01:   Labeling

27.01:   Submission

28.01:   Terminology

5

The Philosophy
Behind the Policies

This chapter provides key questions to increase the awareness for the need of the suggested policies as well as background information on the topics. It also generally introduces the reader to the philosophy behind the policies. Key policies on "policy" and auditing will be discussed first in order to give the general outline of the quality system. The other topics will be presented in alphabetical order. Consideration was given to clustering topics that are closely linked and sometimes overlapping (e.g., documents for regulatory purposes, dossier, electronic dossier, global dossier, and submission; electronic submission, information technology, and tools); however, this was not done, because it would have caused too much repetition. Instead, related policies have been referenced where applicable.

POLICY 01.
THE POLICY ON POLICY

Why Is This Policy Needed?

Do you want everybody to create their own format, structure, and content of policies? Should you go to the trouble of harmonizing documents after they have been generated in an individual style? Do you prefer to settle a few basics before starting? Do you believe you can start with SOPs right away? Then good luck to you. However, if you are willing to listen to reason and experience, you will use a top-down approach to reach general agreement first (policies), then settle on details (standards and Standard Operating Procedures [SOPs]). It will be much easier for everybody.

Policies as Part of the Quality System

The intention and goals of the Quality Manual require that the Manual contains a Policy on Policy, which defines the standard for format and content. For a Regulatory Affairs department, this is done in the Policy on Policy.

 If a company already has a quality system in place, either for the whole company or on a higher hierarchical level than Regulatory Affairs, it may be necessary to adapt the standard for format and content and subsequently all policies. The policy on Auditing and Compliance and the Policy on Policy are the two key policies of the quality system, therefore, they are presented first, with all of the other topics following in alphabetical order.

POLICY 02.
THE POLICY ON AUDITING AND COMPLIANCE

Why Is This Policy Needed?

Just imagine: A Regulatory Body inspects your company and asks to see your procedures. They also want to make sure that you are in compliance with your policies, standards, and SOPs. Without regular auditing, you can never be sure whether your quality system works and whether the necessary improvements take place.

If you intend to use this Quality Manual only to show off and do not intend to make it happen—yet another good idea stopped dead in its tracks—do not read on!

Auditing and Compliance

Regular auditing is a prerequisite for a functioning quality system. After the mission and goals are clear, and the quality of the products and procedures required to achieve these goals has been defined by policies, standards, and SOPs, it is important that the following takes place:

- A Quality Manual containing all policies, standards, and procedures is appropriately maintained and updated.

- Revisions are distributed to the appropriate person(s) in a timely fashion.

- The current version of the Quality Manual is available to each Regulatory Affairs employee during everyday work.

- Regular trainings are conducted to ensure the correct performance of the quality system.

- The Quality Manual is adhered to in everyday work.

- Regular feedback is solicited concerning the need for improvements in quality and modification(s) to the Quality Manual.

Regular auditing of the quality system ensures this by noting/analyzing the actual status and also increasing awareness. Audits should not be perceived as controls but rather as a regular assessment of the status of the quality system. It is also a chance for self-assessment and a development tool for each employee. This is also related to the culture of the company (e.g., whether mistakes are seen as chances for improvement rather than as punishment).

The purpose of an audit should be clearly stated to all department(s)/function(s) prior to any audit. An audit plan should be prepared and distributed in advance. It is equally important that the department(s)/function(s) to be audited are represented during the audit, as well as internal or external customer(s) and supplier(s). It may also be helpful to include somebody to act as neutral third party (e.g., a member of a Contract Research Organization [CRO] or an employee from Corporate Quality Assurance).

Any observations made during the audit should be documented immediately and made known to the audited department(s)/function(s). This avoids omissions and misunderstandings. The department(s)/function(s) being audited may be able to clarify any observations. However, if there is indeed a deficiency, the audited department(s)/function(s) must not be allowed to manipulate or influence the auditor(s). However, if the deficiency is corrected during the audit or measures that will correct it are established, such corrective measures may be acknowledged in the audit report.

From the notes taken during the audit, the auditor(s) prepare the audit report that should contain at least the following information:

- Place
- Date
- Name(s) of auditor(s)
- Unit(s) audited
- Policies/standards/procedures audited
- List of observations and deficiencies
- If applicable, measures already taken to correct deficiencies
- Recommendations for corrective measures
- Signature(s) of members of the audit team

The audit report should be distributed to upper management, the Quality Steering Committee, and the audited unit(s). It should be discussed in detail, especially within the department(s)/function(s) audited, to fully understand the observations and deficiencies and their relevance. Thorough analysis must be made to identify the reasons for the deficiencies and define adequate countermeasures. It should also be considered whether deficiencies are due to a lack of understanding or training on the part of employees.

Selected Reading

Kowal, S.M. 1995. Confidential Audits May Not Always Remain Confidential. *RAPS News* (June): 11–12.

POLICY 03.
THE POLICY ON APPLICATION FOR
CLINICAL TRIAL LICENSE

Why Is This Policy Needed?

Would you like to pay a high fine or even go to jail? Would you like to delay your approvals or not receive marketing authorization(s)? If your answer is yes to these two questions, you do not need this policy. But in all probability, you will need a good lawyer and a psychiatrist!

The failure to obtain the Clinical Trial License simply means that you cannot conduct clinical trials in a respective country. If you start the clinical trials without prior approval, you are conducting illegal experiments on humans. This is a severe crime, which may result in a high fine or even a prison sentence; it will also ruin the image of your company.

Clumsy handling of the procedure or a failure to fulfill the regulatory requirements may delay your marketing authorization. If local studies are required for the marketing authorization in a particular country, a failure to meet the requirements for the clinical trial license may eventually lead to denial of the marketing authorization.

Clinical Trial Authorization Application Procedures

Japan

Specific forms must be used that outline toxicity, pharmacology, and other information on use in foreign countries. If applicable, a comparison with similar medicinal products on the market and a scientific rationale for the planned clinical trial may be necessary. Clinical trial samples should be supplied free of charge, otherwise the reason for charging should be given. Notification to the appropriate division of the Ministry of Health and Welfare (MHW) at least two weeks before the start of the trial must be submitted separately for every clinical trial phase.

A summary should be provided to the clinical investigator containing data on physicochemical properties, toxicity studies, pharmacological actions, pharmacokinetics, and, if applicable, clinical results.

United States

The Food and Drug Administration (FDA) is responsible for issuing clinical trial authorizations and reviews Investigational New Drug (IND) applications. Generally, the following information is required:

- Product and applicant

- Active substance(s)

- Finished dosage form(s)

- Pharmacology, pharmacokinetics, and toxicology in animals

- Pharmacology and pharmacokinetics (if available from foreign studies) in humans

- Foreign clinical data (if available)

- Clinical Trial Protocol(s)

- Institutional Review Board (IRB) approval of each protocol

- Investigators' names, addresses, and qualifications

- Annual reports and Adverse Event Reports

The investigator's brochure should summarize the actual state of knowledge on the product, nonclinical tests, and, if applicable, prior clinical tests relevant to the clinical investigation in question.

European Union (EU)

So far, only national procedures for clinical trial authorization exist in the EU; however, a draft proposal (1) is now available.

Note

1.   III/5778/96, final draft (1/97), Proposal for a Directive . . . of the European Parliament and of the Council of . . . on the approximation of provisions laid down by law, regulation, or administrative action relating to the implementation of Good Clinical Practice in the conduct of clinical trials on medicinal products for human use.

Selected Reading

Fox, T. 1996. The U.S. IND: Practical Aspects. *Regulatory Affairs Journal* 7 (5):371–377.

IFPMA. 1994. *Compendium on Regulation of Pharmaceuticals for Human Use.* Geneva: International Federation of Pharmaceutical Manufacturers Association.

Legrand, C. 1995. Clinical Trial Initiation Procedures in Europe: The Legal Framework and Practical Aspects. *Drug Information Journal* 29:201–259.

Yakugyo Jiho Co., Ltd. 1992. *Drug Approval and Licensing Procedures in Japan.* Tokyo.

Yakugyo Jiho Co., Ltd. 1993. *Supplement to Drug Approval and Licensing Procedures in Japan.* Tokyo.

POLICY 04.
THE POLICY ON APPLICATION
FOR MARKETING AUTHORIZATION

Why Is This Policy Needed?

Would you like to pay a high fine or go to jail? Would you like to risk losing your manufacturing license? If your answer is yes, do not read on, but get a good lawyer and (possibly) a psychiatrist!

The failure to obtain a marketing authorization simply means that you cannot market the medicinal product in the respective country. If you do, you will be accused of selling potentially harmful medicinal products illegally which makes you eligible for paying a high fine or a sentence in jail. This does not even include what will happen to the image of your company. Clumsy handling of the procedure or a failure to fulfill the regulatory requirements may delay your time to market or, as a worst case, will exclude your medicinal product from this market forever.

Procedures for the Marketing Authorization Application

The data requirements of the three regions—EU, Japan, and the United States—are very similar. Typical differences in the marketing authorization procedures are as follows:

- Japan requires the submission of a summary document. The documentation must be on file and available on request. For obvious reasons, (some) trials must be conducted on Japanese subjects.

- The EU uses a top-down approach to data assessment as the basis for the assessment report. The expert opinion and summaries thereby acquire great importance.

- The United States uses a bottom-up approach with reviewers doing their own data evaluation, hence the requirement for the submission of clinical data as SAS files.

Japan

Before submitting an application for a marketing authorization, the applicant may informally contact the MHW to obtain advice. In case of a new chemical entity that is expected to be a major innovation, prior consultation may also be available on clinical trial plans for late phase II and phase III trials.

The submission of the application for a marketing authorization is to the Pharmaceutical Affairs Bureau of the MHW, through a prefectural government. There is a dossier check-in concerning the format and content of the application and supporting data. MHW staff also conducts hearings on the application and, if applicable, inspections of original data and laboratories to assure compliance with Good Laboratory Practice (GLP). Specifications and analytical methods are assessed by the National Institute of Hygienic Sciences or the National Institute of Health. The application is then processed by the Central Pharmaceutical Affairs Council. The applicant is directly informed of the results and may submit additional documents and/or request a hearing. The responsibility for issuing a marketing authorization is with the MHW.

For orphan drugs and life-saving medicinal products, priority review was introduced with the revision of the Pharmaceutical Affairs Law in April 1993.

Although the requirements are the same for foreign manufacturers and Japanese applicants, a local agent is required for foreign manufacturers. Prefectural Authorities may grant marketing authorizations for active substances for pharmaceutical preparations listed in the "Standards of Pharmaceutical Ingredients", and nonprescription drugs (e.g., cold remedies) for which approval standards were established.

United States

The FDA is the body responsible for issuing marketing authorizations. Reviews of the New Drug Application (NDA) are done in parallel by a chemist, pharmacologist, toxicologist, medical officer, biometrician, and biopharmaceutics reviewer. The review results in summaries that are incorporated into the final FDA recommendations. The FDA may also make use of advisory committees. Applicants may request a hearing. The law (Section 505 of the Food, Drug, and Cosmetic Act [FD&C Act]) requires the NDA to be processed within 180 days; however, this deadline is often exceeded. There are also special provisions for orphan drugs under the Orphan Drug Act.

European Union (1)

An outline of and comments on the EU procedures, such as national, decentralized/mutual recognition, and centralized procedures, as well as the community referral are given below.

The EU provides for two separate systems for the issuing of marketing authorizations: National procedures exist in each Member State. They can be followed, if marketing interest is restricted to this market only, or, as will be described below, during a transitional period. However, these national procedures will not be discussed in detail. Guidance on national procedures can be obtained from the national Regulatory Bodies. There are also EU procedures. Guidelines have been issued on the format and content of the applications and the technical operation of these procedures (2).

Some key terms, vital for understanding the procedures, are explained below:

- "Days" usually means calendar days, except where otherwise stated.

- "The person responsible for placing the medicinal product on the market" (3) may be identical to the applicant; sometimes also the license holder or distributor. In internationally operating companies with structures that include corporate, European liaison, and national functions, it may have to be decided on a case-by-case basis who or which institution will act as the "person responsible for placing the product on the market".

- "Identity", referring to the dossier or the summary of product characteristics: Total identity is neither possible, nor does it make sense (e.g., because of differences in language or legal requirements on wording of patient leaflets). Therefore, these passages should be read as calling for identity in content and general structure.

National procedures for issuing marketing authorizations or notification procedures came first. After the establishment of the EU, the concept of a joint market was extended to include medicinal products. In principle, there should be one standard set of requirements applicable in all Member States that allows a medicinal product to be marketed in all Member States of the EU. The Commission believes that all Member States should be able to accept the decision of another Member State on the quality, efficacy, and safety of a medicinal product without additional scientific review, thereby not only guaranteeing free movement of products (once approved) but also cutting costs for both the industry and Regulatory Bodies. These procedures, however, apply only if the product is to be marketed in more than one Member State either by decision of applicant or because of Community interest (see below).

The procedures first created were the so-called multi-state and concertation procedures, precursors of today's decentralized and centralized procedure, respectively. However, mutual recognition did not work out satisfactorily because of differences among the Member States (e.g., political, procedural, medical culture). Industry acceptance of the procedures was minimal because they were too complicated and too slow. Currently, the revised procedures—decentralized and centralized—are being tested. By January 1998, it will be decided whether further revision will be necessary. One option would be to decide that more products will be handled centrally by the European Agency for the Evaluation of Medicinal Products (EMEA) by extending the centralized procedure (or its successor) to more, if not all medicinal products. Presently, there is a proposal to extend the definition of innovative drugs eligible for the centralized procedure, which is a step in this direction. On the other hand, the Commission may conclude that the concept of mutual recognition works satisfactorily and the decentralized procedure should be maintained either in its present form or only as a moderately revised form.

European Agency for the Evaluation of Medicinal Products (EMEA). The EMEA (Figure 1), usually referred to as the Agency, was established by Council Regulation (EEC) 2309/93 (4). Its structure includes a Management Board, a Permanent Secretariat, and Scientific Committees. The EMEA is NOT intended to be an European FDA or super-regulatory body; it should play an administrative and coordinating role. This is also emphasized by the fact

Figure 1. EMEA: Structural Organization

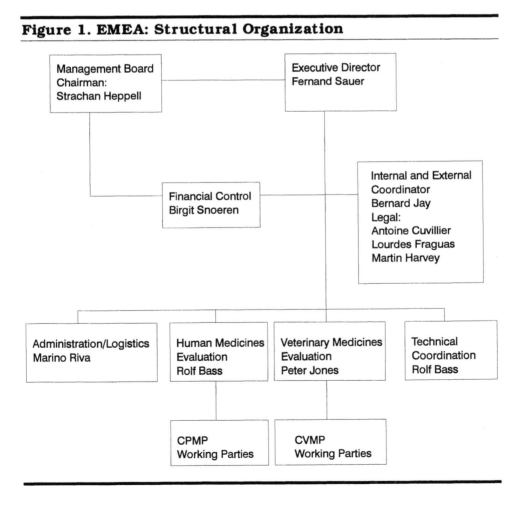

that its Scientific Committees work together with experts from Regulatory Bodies of the Member States during evaluations.

Bearing in mind the reevaluation of the procedures by January 1998, it becomes apparent that not only the procedures—decentralized versus centralized—are competing with one another; the Regulatory Bodies in the Member States are also competing with the EMEA. Time will tell which approach will be the most effective.

National Applications. All Member States have national procedures (Figure 2) in place for issuing marketing authorizations. The requirements for format and content, as well as the technical aspects of the procedures, are set out in national guidelines.

When multiple, parallel national applications for a marketing authorization for a medicinal product are submitted in more than one Member State, the Member State(s) may choose to enter the decentralized procedure (i.e., suspend their own review and await the assessment report of the originating Member State). Until January 1998 (5), they may choose to continue their own national review. After January 1998, however, the decentralized procedure becomes a requirement in situations where multiple, parallel national applications have been submitted. The time frame in which Member States are expected to accept the originating Member State's decision and issue their marketing authorization is 90 days from receipt of the assessment report (6).

After January 1998, national procedures and, consequently, national Regulatory Bodies will lose some of their current importance, as they will no longer review all applications. They will maintain their function (except for the possibility of the rapporteur role in the decentralized procedure) primarily for purely national applications and for homeopathic and similar products (7). (The national route may also remain available for generics if mutual recognition is not possible.)

Decentralized Procedure. The decentralized procedure (Figure 3) is set out by Directives 65/65/EEC (8) and 75/319/EEC (9) as amended for human medicinal products and is applicable for full and abridged applications as well as for variations, provided the original marketing authorization was issued following a decentralized procedure or a multi-state procedure converted to a decentralized procedure (10). It is strongly recommended to in-

Figure 2. National Application/Mutual Recognition

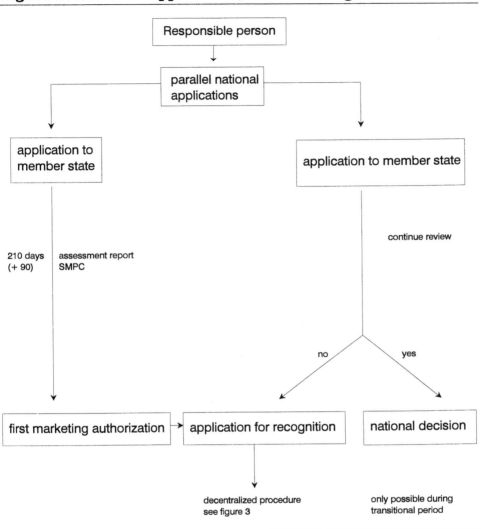

clude all Member States where marketing is planned in a single procedure, although a repeat use of the procedure is possible. However, it should be borne in mind that multiple use of the procedure subjects the original authorization(s) to reassessment with regard to actual requirements and state of the art in therapy, which may lead to a need for additional documentation or reduced claims.

Figure 3. Decentralized Procedure/Community Referral

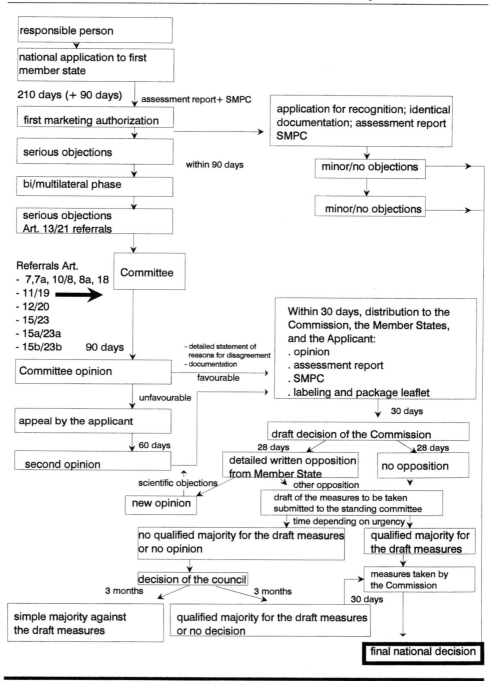

The decentralized procedure is applicable for all medicinal products except

- Products that must undergo the centralized procedure (11)

- Variations to marketing authorizations not issued following decentralized procedure or authorized following a committee opinion before 31 December 1994 (12)

- Medicinal products not yet authorized according to EU standards (13)

- Homeopathic products (14)

The decentralized procedure can be initiated either by a Member State or the person responsible for placing the product on the market.

If multiple, parallel national applications for a marketing authorization for a medicinal product are submitted in more than one Member State, the Member State(s) may choose to enter the decentralized procedure (i.e., suspend their own reviews and await the assessment report of the originating Member State). Until January 1998, they may also choose to continue their own national review (i.e., they are free not to embark into the decentralized procedure). After January 1998, however, the decentralized procedure becomes a requirement in situations where multiple, parallel national applications have been submitted. The time frame in which Member States are expected to accept the originating Member State's decision and issue their national marketing authorization is 90 days from receipt of the assessment report.

The person responsible for placing the product on the market may start the decentralized procedure provided the goal is marketing the medicinal product in at least two Member States. The second (or further) Member State(s) are requested to accept the first Member State's decision by mutual recognition. Of course, the submissions must be identical, especially the summary of product characteristics (otherwise modifications must be indicated). Changes to the documentation may be required to update the documentation to represent the up-to-date status or on request by the first Member State. It may be necessary to resubmit to the rapporteur Member State before the start of the procedure in order for them to review the dossier and become accustomed to it.

The clock starts after all concerned Member States have notified the originating Member State of the receipt of a valid application and the assessment report. The time frame is 90 days in which objections should be clarified between the originating Member State, the concerned Member States, and the applicant. As far as possible, objections should be resolved bilaterally between the originating Member State and the concerned Member State(s). In case of potential risk to human (or animal) health, the concerned Member State(s) must inform the originating Member State, other concerned Member State(s), the applicant, and the Committee. Major objections must be accompanied by detailed reasons and an action plan to correct the application. This is usually completed within 60 days. The applicant may respond in writing or request a hearing in order to resolve the issues. If all objections can be answered satisfactorily, concerned Member State(s) will issue marketing authorizations following mutual recognition within a 90-day period.

The advantage of the decentralized procedure lies primarily in the shorter time to a marketing authorization compared to some of the national procedures, as priority is given by Member State(s) to EU applications over purely national applications. However, it must be remembered that the clock stops during the time that the company compiles the answers to objections. Also, the documentation should be watertight as the procedure tends to uncover most of the deficiencies in the medicinal product. The agreement between concerned Member States tends to apply the sum of each and every relevant requirement. For the industry, this may result in the lowest common denominator regarding labeling claims. It may even happen that the objections are so strong that the marketing authorization issued by the originating Member State must be amended!

For major objections and/or unresolved issues, the matter is referred to the Committee (see below). In the case of deficient applications, there exists the risk of a negative opinion with the consequences of no marketing authorization in the EU at all. This may then have a negative impact on decisions of other countries. The risk of spilling over of decisions is much higher in the case of a negative EU opinion than in the case of a rejection of an application in one Member State only.

Community Referral. It would be a mistake to think of the Community Referral (Figure 3) as just a part of the decentralized procedure for unresolved objections. Listed below are six cases in which the matter MUST be referred to the Committee. Note that it need not be started by the applicant, which means that the Community Referral could be used, for example, by Member State(s) or the Commission, to proactively harmonize marketing authorizations of active substances eligible for reimbursement!

1. Supposed risk to human or animal health (15): concerned Member State

2. Divergent (16) decisions in Member States concerning authorization/suspension/withdrawal, if related to identical dossiers and applications submitted according to Section 4, 4a of 65/65/EEC as amended (17): Member State, Commission, applicant/marketing authorization holder

3. Cases where the interests of the Community are involved (application/suspension/withdrawal/variation) (18): Member State, concerned Member State, applicant/ marketing authorization holder

4. Variations initiated by the responsible person for a product authorized through the decentralized procedure or the Commission's arbitration (19): Applicant/ marketing authorization holder

5. Variations/suspension/withdrawal initiated by a Member State for a product authorized through the decentralized procedure in order to protect human (or animal) health (20) (especially as a result of the evaluation of Adverse Events Reports): Member State

6. No. 4 above and No. 5 above also apply to medicinal products authorized following Committee opinion before 1 January 1995 (21): Member State, applicant/marketing authorization holder

The procedure sets a time frame of 90 days until Committee opinion on behalf of the EMEA, which may be shortened by the Committee in urgent cases. Extension is possible under No. 1 and No. 2 for an additional 90 days. It is important to the industry that the rapporteur in this procedure is appointed by the Committee and is not necessarily from the originating Member State. For negative opinions, the applicant may, within 15 days, request a second opinion under the same procedure. For positive opinions, the draft summary of product characteristics and, if applicable, conditions will be annexed to it as a basis for the draft decision issued within 30 days by the Commission. The applicant again has the right to appeal the opinion within 15 days, but must also submit to the EMEA detailed written grounds for the appeal within 60 days from the receipt of the opinion. Within 30 days, a new assessment will be prepared on which the Committee will decide within 60 days. The applicant has the right to a hearing. Should there still be unresolved objections from other Member States, to be stated within 28 days, the matter is referred back to the EMEA for scientific issues. A final decision will be taken by the standing committee. If there is no qualified majority, the matter is referred to the council for decision within 3 months. The resulting decision is binding for all concerned Member States who will comply with the decision within 30 days.

Centralized Procedure. Council Regulation (EEC) 2309/93 (22) established the centralized procedure (Figure 4) with the purpose of rapidly granting access to important new medicinal products in the entire EU. For biotechnology products (23), the centralized procedure MUST be used, even if the applicant intends to market the medicinal product in only one Member State. For other innovatory products (24), the applicant may decide to use the procedure. An extension of the definition of innovative medicinal products has been suggested with a view to giving more products access to the centralized procedure (25). Variations and renewals subsequently undergo the same procedure. The procedure itself is similar to the decentralized procedure; however, it must be stressed that early contact with the EMEA is of the utmost importance, especially for biotechnology products, to clarify requirements, but also for other innovative products to confirm that the EMEA shares the applicant's view that the product indeed is eligible for centralized procedure.

Figure 4. Centralized Procedure

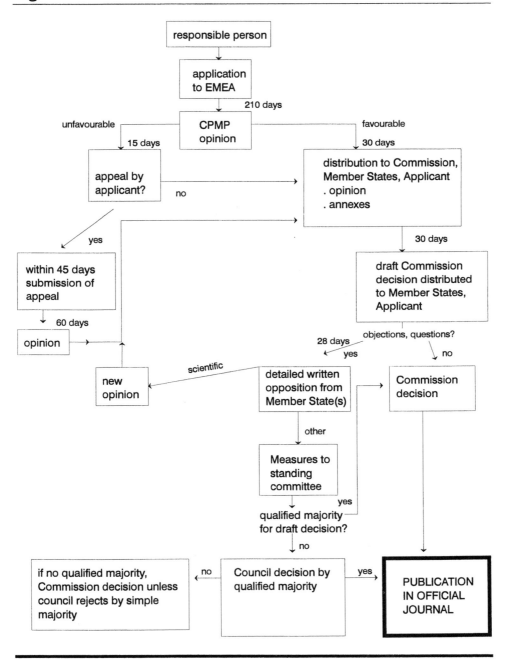

Notes

1. A completely revised and updated chapter based on a previous publication: Dumitriu, H. 1995. Draft Notice to Applicants III/5445/94: What Is New? *Drug Information Journal* 29:1125–1131.

2. III/5944/94: Notice to Applicants for Marketing Authorization for Medicinal Products for Human Use in the European Community (December 1994); III/5429/96, draft (5/96): Notice to Applicants for Marketing Authorization for Medicinal Products for Human Use in the European Union—Revised Version of Parts of Volume IIA; Volume IIB: The Notice to Applicants—Presentation and Content of the Application Dossier, final (1/97).

3. Must be established in the EU, see Art. 2 (EEC) No. 2309/93 (OJ L 214 24.08.93, p. 1) (OJ = *Official Journal of the European Communities*).

4. (EEC) No. 2309/93: (OJ L 214 24.08.93, p. 1).

5. i.e., during the transitional period from 1 January 1995 to 1 January 1998.

6. See III/5447/94 Guideline on the assessment report of 16 November 1994.

7. Art. 14.5, 75/319/EEC as amended.

8. 65/65/EEC: (OJ 022 09.02.65, p. 369). Derogated from 172 B M by 94/1/EGKS, EG (OJ L 001 03.01.94, p. 263); amended by 66/454/EEC (OJ P 144 05.08.66, p. 2658); amended by 75/319/EEC (OJ L 147 09.06.75, p. 13); amended by 83/570/EEC (OJ L 332 28.11.83, p. 1); amended by 87/021/EEC (OJ L 015 17.01.87, p. 36); amended by 89/341/EEC (OJ L 142 25.05.89, p. 11); extended by 89/342/EEC (OJ L 142 25.05.89, p. 14); extended by 89/343/EEC (OJ L 132 25.05.89, p. 16); amended by 92/027/EEC (OJ L 113 30.04.92, p. 8); amended by 93/039/EEC (OJ L 214 24.05.93, p. 22).

9. 75/319/EEC: (OJ L 147 09.06.75, p. 13); amended by 94/1/EGKS, EG (OJ L 001 03.01.94, p. 263); amended by 83/570/EEC (OJ L 332 28.11.83, p. 1); amended by 89/341/EEC (OJ L 142 25.05.89, p. 11); amended by 89/342/EEC (OJ L 142 25.05.89, p. 14), amended by 89/343/EEC (OJ L 142 25.05.89, p. 16); amended by 89/381/EEC (OJ L 181 28.06.89, p. 44), amended by 92/027/EEC (OJ L 113 30.04.92, p. 8), amended by 93/039/EEC (OJ L 214 24.08.93, p. 22).

10. 87/22/EEC: (OJ L 015 17.01.87, p. 38); amended by 93/041/EEC (OJ L 015 17.01.87, p. 38); Art. 4 deals with the Committee opinion; 93/41/EEC repeals 87/22/EEC and contains transitional provisions for applications referred to the CPMP before 1 January 1995 and not having received an opinion before 1 January 1995.

11. Except in the case of applications that were made in accordance with list A (biotechnology products) or B (other "high-tech" products) of Directive 87/22/EEC as amended, for which the CPMP had issued a positive opinion before 1 January 1995.

12. In accordance with Art. 4 of 87/22/EEC.

13. Art. 12 of 75/319/EEC as amended.

14. Art. 14.5 of 75/319/EEC as amended.

15. This refers to quality, safety, and/or efficacy of the medicinal product. Issues not resolved in the decentralized procedure may fall under this heading. See Art. 7, 7a, 10 of 75/319/EEC as amended.

16. Significant divergencies include, for example, a marketing authorization in one Member State versus refusal in another; or different indications or suspension or withdrawal of the marketing authorization(s) based on new data on quality, safety, or efficacy in some but not all concerned Member States. Art. 11 of 75/319/EEC as amended.

17. See Art. 4, 4a of 65/65/EEC as amended. For marketing authorizations granted before 1 January 1995, see transitional regulations. For marketing authorizations issued before 1 January 1995, there were different deadlines for products placed on the market 15 years before notification of Art. 39 of 75/319/EEC as amended: for medicinal products for human use (May 1990) for radiopharmaceuticals, immunologicals, and blood products for human use (89/342/EEC, 89/343/EEC, 89/381/EEC as amended) (December 1992).

18. However, this should not be interpreted as setting up an alternative procedure for new applications; preference must always be given to either the centralized or decentralized procedure. Equally, a referral must be in the interests of the Community and, therefore, must be determined only on a case-by-case basis. For example, an application for a product of high importance from the point of view of protection of human or animal health or the environment (but not eligible for the centralized procedure); suspension/withdrawal because of new data on quality, safety, or efficacy; or a variation because of new pharmacovigilance information.

19. Art. 15 of 75/319/EEC as amended.

20. Art. 15a of 75/319/EEC as amended.

21. Art. 15b of 75/319/EEC as amended.

22. (EEC) No. 2309/93: (OJ L 214 24.08.93, p. 1).

23. Medicinal products developed by one of the following biotech processes: recombinant DNA technology; controlled expression of genes coding for biologically active proteins in prokaryotes and eukaryotes, including transformed mammalian cells, hybridoma, and monoclonal antibody methods.

24. Extract from Annex to Regulation 2309/93:

- Medicinal products developed by other biotechnological processes that, in the opinion of the Agency, constitute a significant innovation.

- Medicinal products administered by means of new delivery systems that, in the opinion of the Agency, constitute a significant innovation.

- Medicinal products presented for an entirely new indication that, in the opinion of the Agency, is of significant therapeutic interest.

- Medicinal products based on radioisotopes that, in the opinion of the Agency, are of significant therapeutic interest.

- New medicinal products derived from human blood or human plasma.

- Medicinal products in which the manufacturing process, in the opinion of the Agency, demonstrates a significant technical advance, such as two-dimensional electrophoresis under microgravity.

- Medicinal products intended for administration to humans that contain a new active substance that, on the date of entry into force of this Regulation, was not authorized by any Member State for use in a medicinal product intended for human use.

25. Van Essche, R., with the key support of Prof. Benzi, Member of European Parliament (MEP). 1996. *European Development of Innovative Drugs.* Paper presented at the RAPS meeting Amsterdam, 22 April 1996.

Extract from Annex 1 on human innovative drugs (evaluation scale):

1. Drugs that show therapeutic efficacy for a disease or a symptom for which there is no active drug available.

2. Drugs that show therapeutic efficacy for a disease or a symptom for which an effective drug is already available but whose effect is necessary for a subset of the affected population.

3. Drugs that are more effective and/or show less serious adverse effects than the reference drug of an equivalent therapeutic effect.

4. Drugs that may be given to special groups of patients with increased efficacy or reduced toxicity.

5. Drugs that are presented in a form that is more practical and/or convenient for the patient.

Selected Reading

IFPMA. 1994. *Compendium on Regulation of Pharmaceuticals for Human Use.* Geneva: International Federation of Pharmaceutical Manufacturers Association.

Yakugyo Jiho Co., Ltd. 1992. *Drug Approval and Licensing Procedures in Japan.* Tokyo.

Yakugyo Jiho Co., Ltd. 1993. *Supplement to Drug Approval and Licensing Procedures in Japan.* Tokyo.

POLICY 05.
THE POLICY ON ARCHIVING MANAGEMENT

Why Is This Policy Needed?

Have you ever lost a marketing authorization document? Your company's management will "love" you for it. Hopefully, the Regulatory Body is willing to issue another one. However, it has happened that Regulatory Bodies have requested a copy from the applicant because they could not find their copy.

How long does it take you to provide copies of all presently used package leaflets of your company's medicinal products? If it takes longer than the time to read this sentence, then it is simply too long if a pharmacovigilance issue is cooking or if pharmaceutical critics are after your company.

Do you know how many cubic metres of archived documents you have? No? Then you are probably wasting your company's money on unnecessary archiving and at the same time running a serious risk of not archiving what should be archived. What would happen if your whole building burned down today? How soon would you be operative again? Any answer except "the next working day" is NOT COMPETITIVE.

Archiving Management

The importance of archiving may not exactly spring to mind when pondering important functions in the medicinal product development and maintenance process. However, it is of the utmost importance for a company to reach a common understanding and a general consensus on the necessity, scope, and responsibilities for archiving.

Archiving presents itself as a two-edged problem: the danger of not fulfilling legal obligations due to incomplete knowledge of all applicable requirements versus the danger of redundant and/or excessive archiving of records.

Pros of Archiving

In order to meet all applicable legal and/or business obligations, an assessment must be carried out by country, business obligation, contract, and record type. Storage requirements should be clearly stated both for original and duplicate, specifying at least the time and medium of archiving. Other requirements might also apply to location and limitation of access. Remember that different legal and/or business obligations may apply to one record type (e.g., drug law, good practice requirements, commercial law, and tax law).

Companies operating in more than one country should assess the possible consequences of the application of one country's legislation in another country(ies). For example, in Germany, any records that might be of relevance in liability suits must be archived for a period of 30 years following the damaging event (1) which means that they must be archived for 30 years after the medicinal product is withdrawn from the German market, even if the company is not based in Germany. Therefore, companies are well advised to meet the highest applicable standards for archiving.

Storage media must also be considered. In a court of law, oral testimony often has precedence over written testimony, which is generally considered hearsay evidence (2). Furthermore, a signed original has precedence over record forms that could have been tampered with more easily (e.g., paper copy or microfiche/image/electronic record). Now considering the long product development times and the frequency with which personnel change jobs, it will often be difficult to have the author of a document testify in court, hence the importance of records and, especially, original documents. An inability to produce an original or master copy (e.g., in case of patent or liability suits) could weaken your position considerably.

Besides legal and/or business obligations, some records may be needed by individuals or departments in their everyday work.

It is important to define customer needs (e.g., frequency of access, access time, and the need for copies).

For some companies, historical interest must also be considered: Records may allow future generations valuable insight into the organization and functions of the company, and into the development of major medicinal products and therapeutic breakthroughs.

Cons of Archiving

Consider the rental costs per cubic metre of archived material as well as personnel and handling costs. Because cost cutting is of the essence today, only those records that are needed in order to fulfill legal and/or business obligations or that are required for everyday work should be archived.

Archived material may also turn against you in court; therefore, the company should have information management guidelines governing the generation and distribution of information as well as the generation of records. The company policy on archiving should state the following:

- Avoid the generation of unnecessary records.

- Limit the storage of records by defining times when records may be destroyed.

- Provide for regular cleanup and reporting.

- State responsibilities clearly.

Notes

1. German Drug Law (including 5th amendment), § 84 in conjunction with § 90.

2. Dixon, R. 1995. Document Image Processing—Ross Dixon Discusses Some Problems and Developments Concerning the Legal Admissibility of DIP in the UK. *Regulatory Affairs Journal* 6 (5):370–375.

POLICY 06.
THE POLICY ON CHANGE ALERT/
AUTHORIZATION PROCESS

Why Is This Policy Needed?

Would you be surprised by what is actually being sold by your company in the marketplace? Are there tablets of different color and shape, leaflets with different labeling from what is known by the Regulatory Bodies? You do not know? You are not sure? Then I strongly recommend that you learn this information or secure the services of a good lawyer. There is an imminent threat in connection with selling unauthorized or mislabeled drugs. This criminal offense could cost you dearly—either a large fine or a sentence in jail, not to mention liability suits. This is what will happen if unauthorized product is put on the market. Do you want anybody, including yourself, to come to harm?

Product Variations

Any product variation should be considered under the following aspects:

- Impact on the quality, safety, and efficacy of the medicinal product

- Impact on the information submitted to Regulatory Bodies

Variations to a medicinal product can consist of changes in the source and synthetic pathway of the active ingredient and excipients, in qualitative or quantitative composition (including packaging), in the manufacturing process, in the specifications, in the analytical methods used, in the stability claim, and in the labeling and product information. Some variations cannot be totally avoided, for example, the product information must be adapted if broader use reveals rare or very rare side effects. However, the majority of variations can be avoided by careful pharmaceutical development and validation of the manufacturing processes as well as adequate establishment of the right dose during clinical testing. Consider how much time and capacity are spent on variations by both the industry and Regulatory Bodies—time and capacity that might be spent better.

Variations are generally undesirable from a regulatory perspective as the dossiers and the submissions for a medicinal product become more complex and thus more difficult to handle. The potential negative consequences of variations are as follows:

- For variations during development: Proof, even bridging studies, may be required to demonstrate the equivalence of dosage forms. The pooling of clinical data may be impossible. There may not be enough patients treated with the dosage form that is intended to be marketed.

- The need to update already submitted information in all applicable countries with considerable capacity and time consumed by Regulatory Affairs and the concerned Regulatory Bodies.

- Differing claims in different countries that may be considered unethical and may make the company vulnerable to pharmaceutical critics. Differing claims may even invite questions from the Regulatory Bodies.

- Different quality standards may result in separate medicinal products and, thus, to the burden of extra costs for separate manufacture.

- Illegal products, if notification or approval of variation is required and was not submitted or applied for. Consequences could include product recall, fines, liability suits, and even loss of the manufacturing license.

Therefore, companies must deploy adequate in-house procedures that guarantee that each and every proposal for a variation is evaluated by Regulatory Affairs. Regulatory Affairs must have the right to veto proposals for variations. In any event, Regulatory Affairs must have the right to specify the conditions that must be fulfilled before the variation is developed and the modified medicinal product is placed on the market.

Selected Reading

Dumitriu, H. 1997. Postapproval Changes in the United States—Is Life Getting Easier for Industry? *Drug Information Journal* 31 (1): 143–149.

Quilliet, A., Ratouis, R., and Scheeren, J. 1996. Variation of Marketing Authorizations. *Regulatory Affairs Journal* 7 (4):288–294.

POLICY 07.
THE POLICY ON CONTACT REPORT

Why Is This Policy Needed?

Do you always understand what the regulators tell you? If so, you should seriously consider a career as a mind-reader. For the rest of us, the only possibility is to make sure that we have really understood each other by writing things down and exchanging notes.

If, during the marketing authorization procedure, the agreements that you thought you had seem to have vanished into thin air and your management does not recall the same things, you will probably wish that you had written a contact report.

Contact Report

Regulatory Affairs is responsible for managing contact with the Regulatory Bodies. When internal or external experts or other representatives of the company wish or are requested to contact the Regulatory Bodies, Regulatory Affairs should arrange the contact and be present during the meetings.

A contact report should be written for each contact. The writing of contact reports is important and should be done carefully in order to

- Fix in writing the results of the contact or any other important information received.

- Make sure that all parties to the contact have the same understanding.

- Distribute to interested parties for information and actions to be taken.

- Provide a document for future reference.

Great care should be taken to adequately represent the results or positions taken during the contact. *Adequately* in this context means *fully*, without omitting important items, with fairness to the contact partners (e.g., no negative comments on the Regulatory Body's representatives or their concerns; not too optimistic, because this might create unrealistic expectations within the organization; not too pessimistic, as this might lead to a loss in reputation).

POLICY 08.
THE POLICY ON CONTACT WITH REGULATORY BODY

Why Is This Policy Needed?

Imagine what might happen if the reviewer during his or her review of your company's medicinal product is contacted in parallel by company management, the experts in your company, the project manager, or Regulatory Affairs—all with the same questions and concerns or contradictory or seemingly contradictory statements. How would you react? Obviously, the reviewer will be somewhat irritated, which will probably result in a less favourable opinion about your medicinal product.

Contact with Regulatory Bodies

Direct contact with the Regulatory Bodies is extremely helpful in order to achieve the goals of Regulatory Affairs (e.g., to reach/ maintain high quality marketing authorizations). Good contacts with the Regulatory Bodies will enable you to communicate the standpoint of the company effectively and to establish a friendly and productive atmosphere. It is also important to prove continuously that the quality standards of the company are adequate in order to let the Regulatory Body build up a certain amount of trust in the company and Regulatory Affairs as its representative. It must be borne in mind that there is often a considerable amount of mutual distrust and lack of information on both sides, which Regulatory Affairs should help to overcome. Industry representatives sometimes regard themselves as the only experts and the regulations primarily as an obstacle to placing the product on the market. Regulators are perceived as bureaucrats far away from practice. On the other hand, regulators sometimes regard themselves as the only body that protects patients from hazardous medicinal products. Thus, they challenge the intentions and objectivity of the industry. Regulatory Affairs should perceive itself as a mediator and interpreter between the Regulatory Bodies and the company. It must give its best efforts to communicate each party's interests and concerns to the other adequately.

Regulatory Affairs is solely responsible for contact with the Regulatory Bodies. If internal or external experts or other representatives of the company wish or are requested to contact the Regulatory Bodies, Regulatory Affairs should arrange the contact and be present during meetings.

Regulatory Affairs is also responsible for maintaining the list of addresses of the relevant Regulatory Bodies, and, if applicable, lists of addresses of the contact partners within the Regulatory Bodies for special medicinal products or groups of medicinal products. Furthermore, Regulatory Affairs is responsible for the technical organization of contacts (e.g., meetings or hearings).

A contact report should be written for each contact. The writing of contact reports is important and should be done carefully in order to

- Fix in writing the results of the contact or any other important information.

- Make sure that all parties to the contact have the same understanding.

- Distribute to interested parties for information and actions to be taken.

- Provide a document for future reference.

Great care should be taken to adequately represent the results or positions taken during the contact. *Adequately* in this context means *fully*, without omitting important items, with fairness to the contact partners (e.g., no negative comments on the Regulatory Body's representatives or their concerns; not too optimistic, because this might create unrealistic expectations within the organization; not too pessimistic, as this might lead to a loss in reputation).

POLICY 09.
THE POLICY ON CRISIS MANAGEMENT

Why Is This Policy Needed?

What would you do if you missed the deadline for the renewal of the marketing authorization for your company's biggest product in its largest market? What if the Regulatory Body simply requests that you to take the product off the market and apply for a new marketing authorization? Don't laugh! This happened to a well-known company in Europe.

What would you do if your whole building/facility burned down? What if this happened in the middle of an application process, for example an EU procedure? What would you do if an important product change requiring prior approval has been already completed and unauthorized product is being marketed worldwide? What would you do if the department head has a heart attack?

These are only a few examples. If your only reaction is, "it has never happened so far and hopefully never will", you will almost certainly be floored when it does happen. And you can bet that it will happen—at the worst time possible!

Some Reflections on Crisis Management

A crisis may be defined as a situation with a high potential of danger or damage to the company's reputation, substances, medicinal products, or personnel. Usually, a crisis will appear unforeseen or suddenly, or a situation that appeared under control will worsen within a short period of time. Typically, a crisis produces both time pressure and emotional pressure. Quick and well-planned action will be required to improve the situation and/or prevent further damage.

Since the mid 1980s, considerable research has been done on the subject of crisis management. Meetings and workshops on this subject are abundant (1) as well as literature. Today, a crisis is viewed not so much as an event but rather as a process, for which the company should assume responsibility (2).

If you analyze past situations of crisis that have developed into catastrophes (e.g., the sinking of the *Titanic*), you will find that, as a rule, they were unforeseen but not unforeseeable. They were not so much an event (the *Titanic* ramming into an iceberg) as a process of several things, where quality standards were not met: If binoculars had been available in the lookout, they might have seen the iceberg in time (binoculars have now been found on board the wreck, carefully locked away). Failing that, if only they had enough lifeboats, the crisis would have never turned into the catastrophe.

A crisis might also be defined as a situation so disagreeable that you would prefer even not to imagine it. This mental block is a normal reaction. However, it is difficult to identify any advantage, except to shield from momentary discomfort. On the other hand, if your mind circles around crises permanently, the danger of self-fulfilling prophecy arises.

From the treatment of anxiety disorders, we find that not thinking about a crisis is not the key to the solution. The same applies to possible crisis situations: Identify and discuss possible crisis situations in order to bring them into perspective and to gain confidence that you will be able to handle them. This will make you fit to deal with truly unexpected crisis situations that cannot even be imagined.

The basic principles that should be followed in case of crisis are as follows:

- Immediately elevate the crisis to the highest applicable decision-making level.

- Isolate the problem and develop solution(s).

- Create a communication plan with a single voice or contact directed to interested parties.

- Drive management to do the right thing.

- Follow up on measures to be taken (3).

Try to perceive crisis situations as events that will occur. Do not overly concentrate on the negative aspects of a crisis. The positive aspects of crisis might lead to a greater maturity of a company, similar in nature to the growing maturity of a person from childhood to adulthood. A crisis can lead to the elimination of counterproductive efforts and may be used as the starting point for radical change in an organization (4).

Notes

1. For example, workshops and meetings organized by Drug Information Association (DIA) and Institute for International Research (IIR). For addresses of these and other organizations offering education and training, see the respective chapter on education and training.

2. Forgues, B. 1996. New Approaches to Crisis Management. *Rev-Fr-Gestion* 108:72–78.

3. McHenry, J. 1996. Panic-Free PR Crisis Management. *Marketing Computers* 16 (2):26–27.

4. Pauchant, T. and Morin, E.M. 1996. Systemic Crisis Management and the Avoidance of Counter-Productivity. *Rev-Fr-Gestion* 108:90–99.

Selected Reading

Albrecht, S. 1996. Crisis Management for Corporate Self-Defence: How to Protect Your Organization in a Crisis; How to Stop a Crisis Before It Starts. New York: American Management Association.

The Business Protection Plan. 1995. Phoenix, AZ: League Publications.

Computing and Communications in the Extreme: Research for Crisis Management and Other Applications. 1996. National Research Council, Computer Science & Technology Board Staff.

Devlin, E.S., Emerson, C.H., Wrobel, L.A. 1994. Business Resumption Planning (includes forms on disks). Boston: Auerbach Publications.

Fearn-Banks, K. 1996. Crisis Communications: A Casebook Approach. Mahwah, NJ: Lawrence Erlbaum Associates.

Gennery, B. 1996. Crisis Management—Brian Gennery Discusses the Methods Used by Regulatory Professionals for Dealing with Regulatory Crises. *Regulatory Affairs Journal* 7 (7): 564–566.

Lerbinger, O. 1996. The Crisis Manager: Facing Risk and Responsibility. Mahwah, NJ.: Lawrence Erlbaum Associates.

Lines. 1995. *Crisis Management.* New York: McGraw-Hill.

Richardson, B. and Smith, D. 1995. *Crisis Management: A Workbook for Managers.* New York: John Wiley & Sons, Inc.

POLICY 10.
THE POLICY ON DOCUMENTS
FOR REGULATORY PURPOSES

Why Is This Policy Needed?

When it comes to regulatory documents, you have two options:

1. You can just pass on everything you receive from the scientific disciplines and leave it to the Regulatory Body to sort it all out. If so, then why do you call yourself Regulatory Affairs? The company post office would be a better name.

2. You can also check each and every document—format, structure, content, and on to typing errors. It would be better to write them yourself. Only, do not expect thanks from anybody.

But there is a compromise between these two options: Tell the scientific disciplines what the Regulatory Bodies expect to see and agree on internal standards. Then let the disciplines do their job and you do yours.

The Standardization of Documents for Regulatory Purposes

Definitions

*   Document for regulatory purposes: Any document that is intended for regulatory purposes (e.g., application for clinical trial authorization or application for marketing authorization).

*   Dossier: A compilation of documents relevant for a specific regulatory purpose (e.g., application for clinical trial authorization or application for marketing authorization), in a specified country(ies) for a developmental or marketed medicinal product in a structured form (i.e., submission-like). If applicable, it is a subset of the global dossier. The dossier is the basis for the submission(s).

*   Global Dossier: A compilation of all documents required for international regulatory purpose(s) for a developmental or already marketed medicinal product. It is maintained continuously throughout the life cycle of the medicinal product and serves as a repository for the generation of dossiers and submissions.

*   Submission: A country-specific compilation of documents for a specific regulatory purpose (e.g., application for clinical trial authorization or application for marketing authorization) for a developmental or marketed medicinal product in a structured form according to national regulatory requirements. It is based on the dossier, or, if applicable, the global dossier. It may contain additional national documents (e.g., national leaflets or application forms).

Regulatory Requirements

Documents for regulatory purposes must meet regulatory requirements. However, as the information contained in documents for regulatory purposes is mainly generated outside Regulatory Affairs by various scientific disciplines, the regulatory requirements can be met only by a joint effort between Regulatory Affairs and the disciplines. The final target is the customer outside the company (i.e., the Regulatory Body). The internal customer of such documents is Regulatory Affairs, who acts as the interpreter of the Regulatory Body's requirements by drawing on experience in Regulatory Affairs. However, the disciplines should interpret the same regulations using their scientific education and experience. This makes sense as it mirrors the way regulations are originally generated within the Regulatory Bodies, namely in collaboration with scientific and administrative regulators.

Documents for regulatory purposes typically must also serve other purposes as well, including the following:

- Good practice (e.g., documents used as operating procedures)

- Company requirements for content and format (e.g., corporate logo)

- Legal/business requirements (e.g., marked as confidential)

It follows that the process of generating documents for regulatory purposes must provide for adequate input from both the scientific disciplines and Regulatory Affairs and also feedback from Regulatory Bodies (e.g., by letters of deficiency). However, for the sake of empowerment, the responsibility for the documents should remain with the individual authors. The approach to the generation of documents for regulatory purposes will vastly depend on the company's organization and the number of products, involved departments, and so on. Generally, there are three possible strategies a company might choose to solve this problem: in one casting, modular approach, or fractal approach.

In One Casting. To make a dossier or a part of it (e.g., the CMC section), in one casting offers the benefits of a harmonized document with a logical structure. The expert, who is writing the expert report and summary, is also the person preparing the whole section. The resulting dossier is usually a very convincing one. However, the drawbacks are that—except in very small companies—the people responsible for manufacturing, carrying out the tests, and so on cannot relate to the document for regulatory purposes. This results in a danger that changes are not adequately reflected in the documentation. Also, the documents and the dossier will be written for a specific regulatory purpose. For submission in other countries, it will have to be rewritten to adapt at least the structure to the regulations of those countries. The process itself involves an immense concentration of know-how in one person. The process is also quite slow, and not many dossiers can be generated in this way at the same time. The approach is quite similar to the creation of a unique masterpiece in contrast to mass production.

The Modular Approach. The modular approach (1) is the other extreme, typically used in very large companies and somewhat similar to the manufacturing technologies that Japanese companies have used successfully. The dossier is broken down into the smallest possible single units, according to regulatory requirements and departmental organization. Each department issues these modular documents under its own responsibility. There is a system in place that informs all departments/authors about changes in regulations or in-house agreements. Usually, a final coordination and harmonization of these documents is required before submission. The benefit of this approach is that the responsibility for the individual documents for regulatory purposes remains with the author. Every type of dossier can be easily generated simultaneously from the modules without having to write country-specific documents. But there is also a drawback: The number of required modules is quite large (300–600). Considerable capacity is required for the maintenance and the follow-up of internal agreements, templates, and so on. It may also be difficult for individuals to relate to the complete submission.

The Fractal Approach. In the automobile industry, there is a trend away from individual steps performed at the assembly line toward the organization of workers in larger groups that manufacture complete subsystems in a more or less complex array of working steps, typically not more than 100, generated in teamwork to create the finished product. Because these fractals still show a clear relationship to the finished product, the responsible team can better relate its work to the finished product. This approach seems to combine the benefits of the assembly line (e.g., standardization and parallel work) with the benefits of craftsmanship (e.g., high motivation and quality work). The fractal approach might be the solution on how to generate documents for regulatory purposes and submissions most effectively. In practice, this would mean that the individual teams would assume the responsibility of creating specific sections of the dossier—one team each for the CMC, preclinical, and clinical sections.

Summary. Based on the above considerations, Regulatory Affairs must ensure that procedures are in place that allow adequate input from Regulatory Affairs on content and format of documents for regulatory purposes. Early input is more effective than after-the-fact criticism. The example of other industries (e.g., automobile or aircraft) show that the most successful way is to agree on specifications with suppliers and abstain from repeating controls.

The ideal situation is to not only inform the disciplines on regulatory requirements but also provide an evaluation. This is where Regulatory Affairs can be of major benefit to the organization. Also, Regulatory Affairs should not follow in the steps of some CROs who ask for 90–100 percent fulfillment of requirements in order to ensure their own success rate. Regulatory Affairs should also consider the company's need to save time and costs. A certain risk of questions from Regulatory Bodies or even rejection of the application in a country may be acceptable if stated clearly to all concerned parties (e.g., the project team and upper management).

Notes

1. The modular concept began to take shape in the 1980s. Selected articles on the modular approach include the following:

   Cartwright, A.C. and Zahn, M. 1995. The Format and Content of a Global Chemical Pharmaceutical Documentation—A Proposal. *Drug Information Journal* 29:1225–1236.

   Dumitriu, H. 1995. The Industry View of International Standardization of Regulatory Dossiers. *Drug Information Journal* 29: 1125–1132.

Gurien, H. 1991. A Module System for the Preparation of International Dossiers, Manufacturing, and Controls (New Chemical Entities). *Drug Information Journal* 25:285–287.

Jackson, D.K., Piasecki, S. and Adornato, F.A. 1989. Regulatory Perspective on Worldwide Marketing Authorization Applications. *Drug Information Journal* 23:81–86.

Margerison, R. 1989. Recommendations for a Truly International Registration Dossier. *Drug Information Journal* 23:417–420.

O'Brien, M. 1989. U.S. and EEC Requirements for Documenting the Stability of the Active Constituent. *Drug Information Journal* 23:411–416.

Ostmann, M. 1996. Standardization of Report Formats for Chemistry Pharmacy Documents. *Drug Information Journal* 30:201–206.

McKenna, K. 1989. An Overview and Comparison of the U.S. and EEC Chemical and Pharmaceutical Requirements for the Marketing Authorization/New Drug Application. *Drug Information Journal* 23:371–377.

McKenna, K. 1989. Working Group 2: The Final Dosage Form—A Model International Registration Dossier. *Drug Information Journal* 23:529–538.

Ramsay, A.G. 1989. Working Group 1: The Active Constituent—A Model International Registration Dossier. *Drug Information Journal* 23:515–528.

Schuermans, V., Raoult, A., Moens, M., Heykants, J., Reyntjens, A., Saelens, R., and van Cauteren, H. 1987. International Drug Registration Efforts. *J. Clin. Pharmacol.* 27:253–259.

POLICY 11.
THE POLICY ON DOSSIER

Why Is This Policy Needed?

Do you want to lose a lot of time and capacity reinventing the wheel every time a dossier is required? Or do you want to make use of experience gained and prepare guidelines that will quicken the process and give you the time for real project-specific fine-tuning?

Dossiers/Submissions

Definitions

Because there is big intercompany and intracompany variability concerning the use of terms like *dossier, submission, standard dossier,* and *global dossier,* definitions will be given first.

- Document for regulatory purposes: Any document that is intended for regulatory purposes (e.g., application for clinical trial authorization or application for marketing authorization).

- Dossier: A compilation of documents relevant for a specific regulatory purpose (e.g., application for clinical trial authorization or application for marketing authorization) in a specified country(ies) for a developmental or marketed medicinal product in a structured form (i.e., submission-like). If applicable, it is a subset of the global dossier. The dossier is the basis for the submission(s).

- Global Dossier: A compilation of all documents required for international regulatory purpose(s) for a developmental or already marketed medicinal product. It is maintained continuously throughout the life cycle of the medicinal product and serves as a repository for the generation of dossiers and submissions.

- Submission: A country-specific compilation of documents for a specific regulatory purpose (e.g., application for clinical trial authorization or application for marketing authorization) for a developmental or marketed medicinal product in a structured form according to national regulatory requirements. It is based on the dossier, or, if applicable, the global dossier. It may contain additional national documents (e.g., national leaflets or application forms).

Goal

One of the major functions of Regulatory Affairs is to apply for and to obtain regulatory approvals (e.g., clinical trial authorization, marketing authorization, and renewal of marketing authorization). Hence the importance of dossier/submission generation. Regulatory Affairs has the primary responsibility for the resulting product, namely, the dossier and/or the submission. The ultimate goal is to receive qualified approval quickly by meeting the necessary requirements only.

Quality Before Time Before Costs!

Quality will be achieved by meeting the necessary requirements through the submission of data sufficient to meet regulatory requirements and to provide assurance on quality, efficacy, and safety of the medicinal product. Time to market is of the essence. Therefore, time will be cut as much as possible during the submission and review phase. It may even be necessary to submit additional data just to speed up the procedure even if it is disputable whether they are really needed. Costs are a factor when defining the necessary requirements. However, quality and time must have precedence.

Quality for a dossier/submission is quality in terms of content, format, and timely finalization. Quality in terms of content must be created by early and continuous input from Regulatory Affairs during the development process via the generation of internal company standards for documents for regulatory purposes and adequate procedures. The actual contents of a global dossier/dossier/submission should be handled by standards and adapted to specific medicinal products/regulatory requirements. If required, the same applies for quality in terms of format.

Responsibilities and workflow for the generation of documents for regulatory purposes, dossiers, global dossiers, and submissions should be clear in advance. Use procedures that are set up as local Standard Operating Procedures (SOPs).

For locally operating companies, there will be no need to create a global dossier. Also, the dossier and submission generation will usually be combined. For internationally operating companies, there may be two opposing interests in the company:

1. To save time in development and later in maintenance by creating, if possible, one single (global) dossier or, if this is not possible, regional dossiers (e.g., a single dossier for the EU).

2. To customize submissions to the local requirements in order to save time for local approval.

These two opposing interests are often mirrored by the existence of a corporate and a local regulatory affairs function. Corporate Regulatory Affairs will be more knowledgeable about the medicinal product and international regulatory requirements and harmonization efforts, while local Regulatory Affairs will be more knowledgeable about specific local requirements. Therefore, it is advisable to have corporate Regulatory Affairs accompany the development process and generate the dossier and to let local Regulatory Affairs generate the specific submission. In this way, it should be possible to reach the best possible compromise.

As the quality requirements for contents are discussed in other policies, the technical and formal aspects of dossier and submission generation will be focussed on here. As the requirements are similar for dossier and global dossier, only the term *dossier* will be used.

Today, as much as 30 percent of the capacity of the Regulatory Affairs departments is required for the generation of dossiers and/or submissions. The reason for this capacity is, besides a possible lack of standardization, the paper-based approach, which involves a lot of manual work. The use of electronic tools will hopefully improve this situation.

Dossier generation may involve the following steps:

*   Generation of a table of contents (TOC)

*   Generation of cover pages for individual parts

*   Compilation of copies of documents for regulatory purposes according to the TOC

*   Pagination*

*   Other imprints (e.g., "confidential", date)*

*   Cross-referencing from the TOC and the summary part to the documentation*

*   Copying from dossier Master File

*   Insertion into binders

---

*May also be done as part of the submission generation

- Insertion of separation pages

- Labeling of binders

Submission generation may involve the following steps:

- Generation of the TOC

- Generation of documents for specific regulatory purposes (e.g., application forms, national labeling, proof of payment, list of samples) and inclusion in the submission

- Pagination**

- Other imprints (e.g., "confidential", date)**

- Cross-referencing from the TOC and the summary part to the documentation**

- Copying from submission Master File

- Insertion into binders

- Labeling of binders

- Insertion of separation pages

- Production of a cover letter

- Addition of any other material (e.g., samples)

Technical Aspects of Quality

Bear in mind that the Regulatory Bodies will expect generally used terminology and format in the TOC. Any other format will make the dossier check-in procedure and location of information during the review more difficult for the Regulatory Body, thereby losing valuable time for clarification. Also, whenever a header or section is not applicable to the specific medicinal product, this should be so stated rather than the header/section being deleted from the TOC. It is also advisable to adopt the numbering system of the Regulatory Bodies as far as possible.

There is no specific requirement for the cover page, but it is useful to use such cover pages for the whole dossier and the main parts, without overdoing it. Cover pages should state the title and number of the section. Some companies also add the name of

---

**May also be done as part of the dossier generation

applicant, the date of application, the name(s) of medicinal product, the dosage form, and the strength. When cover pages are kept as neutral as possible, they can be printed in advance, which greatly facilitates the process and helps to reduce costs. Also, there may be problems encountered with details (e.g., applicants or trade names might be different from country to country). Some companies combine cover pages with separation pages.

The compilation of copies of documents for regulatory purposes is the most time-consuming step when using a paper-based method. The copying process from originals may lead to deficiencies (e.g., missing pages or bad copying quality). This is where electronic tools are a major improvement. Using an optical archive enables the printing of documents in similar quality as the original and directly in the order of the TOC.

The pagination of a paper file can be done by special machinery or directly by specially equipped photocopiers. Ideally, pagination should be done electronically during the compilation process, including other imprints, such as the date. It may be worthwhile to consider paginating each part or even each volume separately instead of the whole dossier to allow parallel work in order to save time.

When using a paper-based method, cross-referencing can be done only after the documentation has been finalized and paginated. When using electronic compilation tools, hyperlinks make it possible to do a lot of work in advance or in parallel, thus cutting the time to submission.

Again, the copying process by itself can lead to deficiencies in the resulting copies of the Master File. At least one copy should be checked in its entirety; in other copies, spot checks should be made.

Good advance planning/organization of material, workflow, and rooms is essential for binder preparation (insertion, separation, and labeling). Considering that a marketing authorization application for a New Chemical Entity (NCE) consists of 300–600 volumes per copy, electronic tools make it possible to process the printout as it is printed.

Last-minute changes should be avoided as much as possible, especially because of the major impact on the TOC, the pagination, the cross-referencing, and the resulting loss of time. Here, too, electronic compilation tools can be helpful. When using electronic tools, the finalization of the dossier/submission can be started immediately after the last document has been received at Regulatory Affairs. The time to submit depends mostly on the speed of the printer.

Presently, time estimates for dossier generation range from one to six months and one month or less for the generation of the submission. By using electronic tools and adequately standardized procedures, it should be possible to submit documentation within one month after the receipt of the last document at Regulatory Affairs.

POLICY 12.
THE POLICY ON EDUCATION/TRAINING

Why Is This Policy Needed?

Usually, it is no problem to obtain education/training, especially specialized training, for newcomers. However, consider life-saving measures. You probably had training at some time during your life. But would you really be prepared to effectively reanimate someone?

How much do you think you will use a Quality Manual that just sits there on a shelf? Probably never—it is just another glossy binder. How much do you think you will use a Quality Manual that is developed by you and all others concerned? In the beginning you will use it, but what about six months from now?

Let us be realistic: Working within a quality system requires continuous training, and you will need to monitor the trainings —both frequency and outcomes.

People Are Our Most Important Resource

Appropriate quality in terms of regulatory affairs work is determined by the quality of the work of PEOPLE. It may be possible to achieve the goals with insufficient equipment if you have the right people. Recruitment, education, and training are of the utmost importance for the success of the Regulatory Affairs department. Some general guidelines are given in the sections below.

Recruitment

Great care should be employed in generating job descriptions and respective requirements profiles and in the maintenance of these documents. The requirements profile for the Regulatory Affairs manager is as follows:

- Degree in life sciences (preferably pharmacy or medical), Dr./Ph.D. level

- At least three years of experience in Regulatory Affairs

- Excellent communication skills

- Team player

- Languages: English, national language (if applicable), other languages as required

When a position is to be filled, do you aim for the best or do you let your competitor grab key personnel? You might consider using personal contacts, head hunters, or newspaper ads, even the Internet, depending on the kind of position involved. Wording should be based on the respective job description/requirements profile and should be carefully chosen in order to attract only the applicable subset. Otherwise, you will receive many inappropriate job applications.

The next step is to assess the applications and determine who is the best candidate. Usually, this is done through an assessment center and/or an interview. In any case, there should be at least two independent interviews of the candidate (more if possible) in order to have a complete evaluation of a prospective candidate. Try to obtain as much information as possible on the candidate. Sources of information might be references from former positions. However, these can never replace the interview and your personal evaluation. When evaluating many candidates, you might develop a questionnaire for the sake of conformity and later comparison. Use open-ended questions in order to obtain a maximum of information. Questions you should ask in the interview include the following:

- Which assignments in your former position did you like best?

- What do you know about our company?

- Why do you wish to work for us specifically?

- Which aspect of your work do you consider fun?

- What is your professional goal in the next two years?

- What do you consider as your personal assets?

- Which role do you allocate to your family?

- How do you see the future of our area of business?

- Which aspect of the job offered by us attracts you most?

- What do you expect from us?

You should give the applicant a clear idea of the job and the requirements profile. Be honest about the negative aspects of the position.

Education

The previous education of a Regulatory Affairs professional will vastly determine his or her further education and/or training needs. Preferably, the Regulatory Affairs manager's background should be a degree in life sciences at the Ph.D. level or an equivalent. Experience has shown that the broad education of pharmacists provides an easy entrée to Regulatory Affairs, as they have a good working knowledge of chemistry and pharmaceutical development, pharmacokinetics and pharmacodynamics, medical and medicinal information, and, to some extent, regulations. A medical background would also be acceptable because medical knowledge is most difficult to acquire. Biologists and chemists would be acceptable provided they acquire medical knowledge. Many good Regulatory Affairs professionals have varied backgrounds and have succeeded in establishing a career for themselves. However, the trend of the future is away from learning by doing and toward a professionalization through university courses and degrees in Regulatory Affairs. At the end of this policy, various organizations and schools that provide these courses are listed.

Training

Training can be viewed as consisting of six learning objectives: Knowledge, comprehension, application, analysis, synthesis, and evaluation. Employee training programs and continuous learning are a necessity today because the half-life of information is generally one to two years! This is especially true in the fast-moving field of regulatory affairs. The expression "trying to hit a moving target" still applies. The Regulatory Affairs professional must strive to maintain a high degree of knowledge and skills in specialist know-how concerning Regulatory Bodies, procedures, and regulations, as well as in the areas of quality management and electronic data processing.

Training should be planned in advance with employee input and should be closely monitored. It should be viewed as an ongoing process for which the individual employee, the department head, and the personnel department bear responsibility. Training should be documented and feedback should be given.

Long-Term Career

The company/department heads should have clear concepts and ideas of how they perceive the long-term career opportunities of employees and should regularly discuss them with the employees (at least once per year). In the beginning of regulatory affairs, it was the rule that people stayed in the same position until the end of their working life (unless they became head of Regulatory Affairs). This is not acceptable today as the only solutions for employees would be to give up hope of career advancement or else leave the company. Both would result in a loss in motivation and know-how for the company. In the long-term it would lead to a negative selection during recruitment, while the best go to work for the competitors. Leadership today involves identifying and helping potentials. Regulatory affairs is THE department in a pharmaceutical company, offering unique insight into the whole development process and working together with all of the involved departments. Depending on personal assets, the next step might be a specialist career (e.g., leading a local Regulatory Affairs group or a documentation/technical writing department) or a position in project management in order to begin a general management career.

Training/Education Goals

Knowledge and Motivation. Quality work depends not only on know-how but also on the motivation of employees. Therefore, both should be considered when planning education/training measures. It should also be kept in mind that knowledge and motivation do not originate from single trainings but are the result of the work situation in Regulatory Affairs. Department head(s) are responsible for creating an appropriate working climate and serve as an example.

Corporate Identity. Employees must be knowledgeable about the company and its Regulatory Affairs department in order to perform high quality work. Education and training must, therefore, include information on organization, substance(s) and/or medicinal product(s), mission(s), short-term and long-term goal(s), competitor(s), the political environment, and development(s) relevant to the industry.

Social Skills. The Regulatory Affairs manager typically works on teams (e.g., project teams, Regulatory Affairs teams, and industry association working parties), making things happen by mutual agreement in a nice working atmosphere rather than by exercising authority. Contacts with Regulatory Bodies require diplomacy for the establishment of rapport. Therefore, regular education and training will be required with regard to communication, cooperation, and leadership. Conflict management, negotiation skills, rhetorics, and presentation techniques may also be considered.

Quality. In order to maintain a quality system, it is mandatory that regular trainings on all policies, standards, and SOPs are planned and monitored. Documentation of these trainings and feedback must be part of the system. Keep in mind that it is by maintaining/improving the quality system that the organization is learning. Learning is a manageable process, just like manufacturing or marketing.

Specialized Education/Training. Specialized education/training measures and/or certification programs are offered by Regulatory Affairs professional societies and several industry or for-profit organizations. Addresses for information on education/training programs (not a complete list) are as follows:

British Institute of Regulatory Affairs (BIRA)
7 Heron Quays
Marsh Wall
London E14 9XN
Tel: +44 171 538 9502
Fax: +44 171 515 7836

Drug Information Association (DIA)
321 Norristown Rd., Suite 225
Ambler, PA 19002-2755
Tel: +1 215 628 2288
Fax: +1 215 641 1229

or

Postfach
4012 Basel, Switzerland
Tel: + 41 61 382 9019
Fax: + 41 61 382 9050

European Society of Regulatory Affairs (ESRA) (see BIRA)

Central European Society of Regulatory Affairs
(MEGRA—Mitteleuropäische Gesellschaft
für regulatorische Angelegenheiten e.V.)
Megra Seminars
P.O. Box 190325
D-60090 Frankfurt/M, Germany
Tel: +49 69 9738 2024
Fax: +49 69 9738 2033

Institute of Applied Technology, Ltd.
Division of Interpharm International, Ltd.
P.O. Box 2115
L-1021 Luxembourg
Luxembourg

Institute for International Research (IIR)
708 Third Avenue
4th floor
New York, NY 10017
Tel: +1 212 661 8740
Fax: +1 212 661 6677

International Business Communications (IBC)
UK Conferences
Gilmoora House
57-61 Mortimer Street
London W1N 8JX
Tel: +44 171 637 4383
Fax: +44 171 631 3214

Management Forum
48 Woodbridge Road
Guildford, Surrey GU1 4RJ
Tel: +44 1483 570099
Fax: +44 1483 36424

Regulatory Affairs Professionals Society (RAPS)
12300 Twinbrook Parkway
Suite 350
Rockville, MD 20852-1606
Tel: 1 301 770 2920
Fax: 1 301 770 2924

or

15, Boulevard St. Michel
B-1040 Brussels, Belgium
Tel: +32 2 743 15 41
Fax: +32 2 743 15 50

Rostrum Personal Development
Mildmay House
St. Edwards Court
London Road
Romford, Essex RM7 9QD
Tel: +44 1708 776016
Fax: +44 1708 734876

The Center for Professional Advancement
P.O. Box 1052
144 Tices Lane
East Brunswick, NJ 08816-1052
Tel: +1 908 238 1600
Fax: +1 908 238 9113

University Courses

Temple University, Pennsylvania, U.S.: Masters of Science in Drug Regulatory Affairs (requires 10 courses and the submission of a thesis)

Long Island University, New York, U.S.: Masters of Science in Pharmaceutics (with an emphasis on regulatory affairs and quality assurance; requires 10 courses and the submission of a thesis)

POLICY 13.
THE POLICY ON ELECTRONIC SUBMISSION

Why Is This Policy Needed?

Do you realize that an electronic submission could delay your application for marketing authorization by as much as half a year? On the other hand, it just might put your application into the top priority group.

Some Thoughts on CANDA and Other Electronic Applications

Definitions and Scope

CANDA seems to have become a household word with management. However, when asked to submit a CANDA for one of your company's medicinal products, your first question should be: What exactly is meant by CANDA?

CAMA (1), CANDA (2), CAPLA (3), and CARS (4) have been used to signify electronic submissions for regulatory purposes. However, it should be clarified on a case-by-case basis what actually will be the electronic part of the submission. This can range from a few text files of some parts to the completely electronic submission of the dossier, including databases using imaging technology and hyperlinks.

Text Files. Submissions that contain some parts of the dossier on a diskette in word processing format or in ASCII format are often considered to be electronic submissions. Therefore, any statistics on CANDA submissions should be looked at with great caution.

The parts frequently required (e.g., in European countries) in a word processing format are as follows:

- The application form(s)

- The summary of product characteristics (SMPC)

- The proposed labeling (i.e., texts for the patient leaflet, professional information, wording on the packaging)

- The summary part(s) of the dossier (i.e., the expert reports and the actual summaries)

Most frequently, the Regulatory Bodies will ask for submission in Microsoft Word®; the Food and Drug Administration (FDA) used to prefer WordPerfect®, less often, ASCII is required. The Regulatory Bodies should be contacted in advance to inquire about their preferred word processing format.

The reason for this requirement is that part of the information is required for processing the application within the Regulatory Body, thus, it can be directly transferred to the databases without any retyping, thus speeding up the process within the Regulatory Body. In addition, during the assessment of the submission, the Regulatory Bodies produce their own documents (e.g., SMPC and assessment report) based on the texts submitted by the applicant. Making the texts available in a word processing format speeds up the process as only the required modifications need to be carried out instead of reinputting the whole texts.

Making the text available in an electronic format should present no problem to the company; in most cases, the texts will have already been generated with word processing systems. However, it must be kept in mind that most of these systems are not validated and might lead to unintended modifications of the texts. Therefore, the paper copy of the original should always be submitted as the primary reference for regulatory purposes in case of doubt or legal issues. Also, beware of using word processing systems as the sole archival tools for documents for regulatory purposes!

Adding Images and Cross-References. Imaging technology is now being used for electronic submissions (e.g., the DAMOS [5] standard). In addition to the submission in paper format, the whole dossier is made available as images on an optical disc. Cross-references from the TOC to individual documents and from the summary part to the documentation help the reviewer to access the required information quickly. Obviously, it must be ensured that the reviewer is in possession of the appropriate hardware and/or software in order to be able to access the information contained in the electronic submission. At a minimum, the software should be supplied by the applicant. The FDA also accepts specializing training together with loan of hardware and/or software by the applicant.

The electronic image may be the primary reference in case of doubt or legal issues only if the procedure by which the images are derived is validated. However, it should be specified for each submission whether the paper copy or the electronic images are to be the primary reference.

The reason for using imaging technology stems from archival considerations. Over time, dossiers have become bigger and bigger. Today's marketing authorization application for an NCE comprises about 500 volumes, which must be submitted in multiple copies to each Regulatory Body. Management, review, and archiving of such a dossier presents problems for companies and Regulatory Bodies alike if done in paper; the electronic dossier, however, can be easily archived on a few optical discs.

In order to navigate through the dossier, hyperlinks/cross-references will be important as well as the options to access individual pages by page number. Because cross-referencing must be done manually, it is the most time-consuming step in the preparation of this type of electronic submission. However, it must be conceded that some of the cross-referencing needs to be done on paper as well (e.g., for EU applications). If the dossier is a printout of the electronic submission, there is not much additional work. If, on the contrary, a paper submission is used to produce the electronic submission, the referencing is double work. Some other problems may be encountered, for example, the identity of page numbers with that of the electronic submission.

Though the benefit to Regulatory Bodies is obvious, electronic submissions are also welcomed by the dossier check-in function and the administrative staff. For reviewers, the image technology probably does not provide any significant advantages. Reading and reviewing on screen—even a big one—is no fun. Some may enjoy additional gimmicks, such as optical character recognition (OCR) or the ability to produce electronic annotations, but it is the author's feeling that most reviewers browse through the electronic submission to obtain an overview and then start their essential work by accessing the paper documentation.

Adding Databases. For a long time, the FDA has been requesting SAS® files, and any application containing these data is automatically considered a CANDA by the FDA. The reason for this request is that FDA reviewers use a bottom-up approach and insist on doing their own evaluation of the data. The EU, on the contrary, favours a top-down approach: Reviewers are obliged to start with the expert reports and the summaries/tabulated study report and access the documentation only in cases of doubt. The submission of case record forms or databases is thus not required in the EU. A check of the validity of the sponsor's evaluation is considered to be sufficient. However, companies negotiating electronic submissions within the EU frequently encounter individual reviewers who wish to receive databases for the sake of performing their own queries.

Great care must be taken to ensure that all pertinent data, but only that, is made available in an appropriate format. Ensuring this and checking the database by doing some queries of your own may take some time, unless your company uses such software on a regular basis.

Requirements

Regulatory Bodies. Guidance on electronic submissions is available from a number of Regulatory Bodies (e.g., in Germany and in the United States). In the United States, CANDAs are mandatory today for New Drug Applications (NDAs). However, it is advisable to contact the Regulatory Body early when considering an electronic submission in order to know the actual requirements of its reviewers.

Experiments, Projects. The aspect of adverse drug reaction (ADR) reporting is predominant for Regulatory Bodies when considering electronic tools. A number of electronic ADR reporting initiatives exist, such as ADROIT (6), AEGIS (7), Euroscape (8), ICH Working Group E2b (9), INTDIS (10), CIOMS (11), MEDDRA (12), and AERS (13). Active submissions/communication initiatives include the following: CANDA/CAPLA, DAMOS, DDM (14), and ECPHIN (15). Other electronic submission–related initiatives are MANSEV (16), MERS (17), SEDAMM (18), and SMART (19).

Standardization. So far, no electronic standard has been agreed on. (Even though DAMOS seems to be close to becoming the European standard, acceptance in the United States is slow.) As this has been identified as an issue for the International Conference on Harmonisation (ICH), hopefully there soon will be a standard agreed on, because standardization is a must for the use of electronic submissions or in lieu of paper applications.

Who Is the Customer?

The above considerations have been presented in order to update you on history and the present status. However, instead of asking what is exactly meant by CANDA, which in most cases will reveal only pitiful ignorance on all aspects of a CANDA, you should ask: Why should we do a CANDA? Who is it good for? This question is not silly, considering the many ways in which companies have been using CANDAs, namely to produce submissions, to compile the dossier, to track documents, to manage processes, or to maintain documentation throughout the life cycle of the medicinal product.

The customer need not only be the Regulatory Body. Companies using electronic tools might benefit even if they never make an electronic submission to a Regulatory Body. Though the decision to use electronic submission is triggered by the submission aspect, this author believes that the biggest benefits are to be gained in-house by streamlining and reengineering the documentation process, which decreases the time to submission and increases the quality of the submission with regard to format and content.

Submission-Oriented Electronic Submissions. Electronic submissions used to be submission oriented, and many companies still tend to perceive them as something that is created because the regulators want it—just another requirement to be followed. Thus, companies create a paper submission first and then add the electronic part, whatever that may be. In the case of text files, this does not consume much time. However, for dossiers of hundreds of volumes that must be scanned and then hyperlinks and/or cross-references added and the whole thing quality controlled, between one and six months will be added to your time to submit. What Regulatory Affairs basically needs to do in order to obtain a high quality submission is to mimic the review process at the Regulatory Body. This will help you identify mistakes, missing items, and inconsistencies. As a rule of thumb, the electronic submission provides greater transparency. The benefits to the Regulatory Body lie in the areas of handling, dossier check-in control, archiving, and browsing; the issuance of documents, such as the assessment report and the SMPC; if databases are submitted, also query. Review time will not be shortened markedly (unless the submission is of higher quality because of the company's internal electronic review).

Dossier-Oriented Electronic Submissions. Some companies are beginning to use their electronic tools earlier, namely by scanning individual documents for regulatory purposes as soon as they become available in the final version and organizing the images to produce the submission. In this way, some parts may be finalized early and the time to submit will be shortened considerably because referencing and quality checks can be done sooner. Ideally, submission should be possible immediately after the last document is received at Regulatory Affairs. The paper dossier will simply be a printout of what has been generated electronically.

Project Management–Oriented Electronic Submissions. From a project management point of view, the finalization of a document for regulatory purposes is the last step of a work package (e.g., for single-dose toxicity, this is the issuance of the study report). Extending the electronic tool to provide for the tracking of documents and work packages also makes it useful not only to Regulatory Affairs but also to Project Management and the respective scientific disciplines.

Process-Oriented Electronic Submissions. From here it is but a step to demanding an electronic system that manages the whole project documentation process, including the finalization of study reports, document flow, and the compilation of complete submissions. This virtual dossier allows all participants in the medicinal product development process to relate their work to the completion of the final submission. This is highly motivating. Development time can be shortened considerably as loops or possible problems are identified at an early point in time. The system provides a different look at your processes, as some things are made possible by EDP that were not possible before (e.g., simultaneous review). Therefore, new and simpler ways of doing things will be found.

Maintenance-Oriented Electronic Submissions. The most refined application would also handle all updates for all medicinal products during their entire life cycle. Obviously, this would prolong all of the benefits mentioned previously even further and would make life easier for the company concerning variations, renewals, and so on.

Functionality

It should now be apparent that every company may need different functionalities with regard to electronic tools. This section focusses on archiving, compilation, tracking, and review. Document management, of course, is an issue by itself, but this will not be discussed here as there are plenty of publications available on this subject.

Archiving. Imaging technology has been developed explicitly for archiving purposes. If the process is validated, there seem to be no objections to the use of an optical archive. However, as no experience is available as to long-term storage, it is suggested to keep the paper originals until this issue is resolved. Migration to more advanced systems will eventually become necessary as technology evolves (as it did from film to video, cassette to compact disc) but should be feasible. Today, in addition to optical archives (images), truly electronic archives (text, databases) are being established.

Both types of electronic archiving offer easy access to documents, depending at least partly on the quality of the bibliographical information and the indexing. The latter means additional initial work. However, bear in mind that the maintenance of a large paper archive makes some kind of indexing obligatory as well.

When electronically archiving not only submissions after the fact, but single documents for regulatory purposes as soon as they become available, the electronic archive becomes a valuable repository from which—providing the required compilation tools are available—dossiers can be easily generated.

Compilation. Once the images are available—either from an optical archive or scanned in for a specific submission—a tool that compiles the dossier on screen and amends the dossier, if necessary, becomes very valuable. Remember about 30 percent of the capacity of the Regulatory Affairs department is consumed by dossier compilation when using the paper format. Amendments to a paper dossier are capacity- and time-consuming, more so if changes in pagination and cross-references are involved. Keeping the dossier virtual until it is ready for submission makes life so much easier for the Regulatory Affairs professional. Under these conditions, the time to submission is only limited by the printer's capacity once the last document has been received by Regulatory Affairs.

Tracking. Tracking documents in-house, in dossiers, or after submission is one of the most time-consuming tasks of the Regulatory Affairs professional. It is vital that modifications are appropriately distributed. After marketing authorizations have been issued, the registration status becomes more and more complex and electronic tools become a must. So why not use them at an earlier stage of medicinal product development?

Review. The completely electronic submission with its images and cross-references, and, if applicable, databases, is much more transparent than a big paper dossier. It presents a unique opportunity to cross-check the documentation. You can simulate the review of the Regulatory Body to some extent: by simply following through the cross-references, you can identify inconsistencies and missing information. Most CANDA surveys have revealed this function to be a major benefit to companies and, in consequence, to the Regulatory Bodies.

Who Should Be the Supplier of the Electronic Submission?

Determining the supplier of a company's electronic submission is a key decision. On the one hand, this may require specialized knowledge of EDP; if carried out additionally to the generation of the paper submission, it costs additional time and capacity. On the other hand, generating the electronic submission provides an unique experience and the opportunity to have a fresh look on processes and on the documentation. Furthermore, frequent interaction with the Regulatory Body before submission and during review may become necessary, which offers the opportunity to contact on a different level and footing than normally achievable.

Regulatory Affairs. Electronic submissions should preferably be generated in-house, possibly with support from the Information Technology department or the vendor of the system, in order to gain and maintain maximum knowledge and experience in-house. Also, Regulatory Affairs can benefit most from frequent contact with reviewers.

Abstain from doing the electronic submission in-house only in the following cases: If you are under time pressure or it is your first electronic submission. In these cases leave it to a CRO specializing in this type of submission, but try to learn from them as much as possible in order to be better prepared next time.

The In-House Information Technology Specialists. In-house information and technology specialists may be deployed if you have a submission-oriented approach (i.e., a finished paper dossier is converted to electronic submission). However, the beneficial contact between Regulatory Affairs and reviewers during the generation and review of the electronic submission may be lost when selecting this approach.

Contract Research Organization. If it is your first electronic submission or if you are under time pressure, this author strongly recommends that you leave generation of the electronic submission to a CRO specializing in this type of submission. However, this should not become your standard procedure, considering how much experience the CRO is gaining over the years, while you are paying for it.

The EDP Solution

When all previously covered aspects have been given thorough consideration, it is time to look for an appropriate EDP tool. There are three options: Buy, buy and adapt, or develop your own tools.

Buy. Buying is the best solution. It should always be chosen provided there is a product that offers at least 60 percent of your required functionalities. It is the cheapest solution because you have no development to do; you also benefit from the experience of others (development, feedback of other users) by receiving updates, you can use it at once, training will be available and there will be a hot line. There are good solutions available concerning document management, archiving, and retrieval/review. The news is not all good however. The compilation and tracking systems available today are not quite what we would like them to be. Yet this author doubts whether developing your own solution in this area makes much sense because these tools should be up to expectations soon.

Buy and Adapt. When working on the big solution (i.e., integrating the different tools), most companies up to now follow the buy and adapt route. Small-scale development or adaptation is done with the vendor(s) to tailor the system to the company's needs. However, it should be realized that even the smallest project of this kind will take a lot of time and capacity from Regulatory Affairs, initially for defining the specifications and later on for testing and implementing. Do not kid yourself into believing that the resulting system will be more user-friendly than any solution that can be purchased on the market: You are lucky if 60 percent of your requirements are fulfilled satisfactorily. Always remember that Regulatory Affairs personnel, though highly skilled in their job, are not software geniuses trained in programming. You cannot compete with the people working on commercial information systems, including the large number of frustrated users of the first version!

Develop Your Own System. Developing your own tools is a big decision that must not be taken lightly. This author is unsure about recommending it even for very large companies trying to handle their processes as a whole by a single electronic system. The reason is that EDP solutions are very short-lived. They tend to be outdated in six months to one year. Thus, you plan a system now; need at least two years to realize it; and what you get at enormous costs in terms of money, capacity, and time is an outdated system. When it finally becomes productive, you can probably buy superior solutions at every corner. The only reason to attempt this would be that your company is the only one who requires something just like it (which makes one wonder whether it is really needed at all) and that nothing like it is in the foreseeable future. Of course, information technology specialists love it, because it will keep them occupied for the rest of their lives.

Assessment of Risks and Benefits

The evaluation of risks and benefits must be done by each company and also on a submission-by-submission basis. Generally speaking, this author recommends that you try to gain at least some experience with electronic submissions. However, do not engage yourself in the development of information technology systems. Concerning electronic submissions, make sure that there is added value to the company (i.e., that reviewers really are interested in receiving an electronic submission and that it will be reviewed at least as quickly as paper submissions).

Notes

1. Computer-Assisted Marketing Authorization: an universal term used for any type of electronic submission; see also CANDA.

2. Computer-Assisted New Drug Application: the most frequently used, universal term for the electronic submission for marketing authorization, originating in the United States.

3. Computer-Assisted Product License Application.

4. Computer-Assisted Regulatory Submission: a synonym for CAMA or CANDA.

5. Drug Application Methodology with Optical Storage (German BfArM, UK Medicines Control Agency [MCA], and selected pharmaceutical companies).

6. Adverse Drug Reaction On-line Information Tracking System (MCA).

7. ADROIT Electronically Generated Information Service (MCA and selected pharmaceutical companies).

8. Regulatory Bodies of Spain, France, and the United Kingdom.

9. Regulatory Bodies of the EU, Japan, the United States, and industry associations.

10. International database accessible via SWEDIS (WHO; maintained by PharmaSoft AB, Sweden).

11. Council for the International Organization of Medicinal Science.

12. Medical Dictionary for Drug Regulatory Affairs (MCA, CIOMS, FDA, ICH Working Group M1).

13. Adverse Reaction System (FDA).

14. Drug Dossier Manager (Sweden).

15. Electronic communication of pharmaceutical information (now EudraMat) (EMEA).

16. Market Authorization by Network Submission and Evaluation (EU).

17. Multiagency Electronic Regulatory Submission Project (headed by Canada).

18. Soumission Electronique des Dossiers d'Autorisation de Mise sur le Marché (France).

19. Submission Management and Review Tracking (United States).

Selected Reading

Computer-Assisted New Drug Applications. 1995. *Regulatory Affairs Journal* 6 (4):339–340.

Goodman, N.G. 1995. A Brave New World: The Path Toward a Global NDA. *Pharmaceutical Executive* 3:62, 64, 66, 68, 70, 72.

Kirk, S. 1996. The Value of CANDAs in Drug Development. *Drug Information Journal* 30:83–88.

POLICY 14.
THE POLICY ON ENVIRONMENTAL PROTECTION

Why Is This Policy Needed?

How sensitive are you about the environment? The younger generation has been brought up to care about environmental protection, and certainly your company has some statement on environmental protection in one of its glossy brochures. Environmental protection just might save you a lot of money.

Environmental Protection

Some people may be disappointed to find that this section does not deal with regulatory requirements but with the contribution of Regulatory Affairs departments to the conservation of the environment. Indeed, environmental issues must be addressed within Regulatory Affairs today. There have already been instances where Regulatory Bodies have renounced on receipt of additional copies or even certain parts of the dossier (e.g., literature). The German Regulatory Body has started to fight against the flood of plastic wallets for documents.

Common sense should rule the utilization of energy. Waste production should be avoided wherever possible in favor of waste separation and recycling. If the company does not already have a policy or general guidance on environmental protection, Regulatory Affairs should ask for the creation and implementation of such a policy. Especially in bigger departments, an environmental working party should be created with the following responsibilities:

- Conceptual work

  - Concept/plan for waste disposal

  - Plan and implement measures for improvement in waste reduction and recycling (e.g., use of long-lived products)

  - Measures to avoid noise (e.g., use of low-noise equipment)

  - Development of evaluation criteria for ecologic purchase

  - Input and advice concerning plans or measures with possible environmental impact (investments, renovation)

- Everyday work

  - Monitoring of waste disposal

  - Carrying out inspection and analysis of weak spots

  - Implementation and monitoring of legal requirements (e.g., workplace safety)

  - Organisation and carrying out of education/training for Regulatory Affairs employees

  - Continuous advice to Regulatory Affairs employees with regard to environmental protection

  - Close cooperation with the company's environmental health officer

Motivation is very important; therefore, introductory courses should be held to acquaint newcomers with the concepts and principles of environmental protection. Continuous training will be required to maintain motivation. In larger departments, a responsible coordinator and contact partner for environmental protection should be appointed. Try to involve as many Regulatory Affairs employees as possible in the planning of environmental protection measures. Individuals who have been part of the planning and implementation of an environmental protection measure will be more likely to adhere to it and make sure others also adhere to it. Involve practical persons to ensure that measures are realistic. Apart from the environmental working party suggested above, you might also consider competitions for the best idea to improve environmental protection and videos for information and training.

Common sense should rule the utilization of energy. Waste production should be avoided whenever possible in favor of waste separation on recycling.

General Guidelines and Ideas on
Environmental Protection (1)

Purchase. When considering a revision of your ordering system, preference should be given to environmentally conscious manufacturers using the following criteria:

- Less packaging (no unnecessary packaging)

- Packaging made of recyclable material (no mixed plastic materials)

- No Polystyrene®, Styrofoam®

- No polyvinylchloride (PVC)

- Ecologic product design (e.g., small volume/weight, environmentally friendly materials)

- No promotional gifts

- Exclusive use of recycling paper

- Environmental protection part of company strategy

Workplace. Materials used should be free of noxious substances and effects as much as possible. Consider also humane workplace design (e.g., protection from stress through absence of natural light, harmful artificial light, noise, and unergonomic seats). Today there exists a whole range of manufacturers specializing in environmentally friendly alternative products. Before ordering, obtain information on composition (and, if applicable, also manufacture)!

Paper. The work of regulatory affairs consumes enormous amounts of paper. About 80 percent of the waste of a Regulatory Affairs department consists of paper. Did you know that for one ton of paper two trees have to be cut and 120,000 litres of water are needed? For one ton of recycled paper, no trees are cut and only about one tenth of the water is needed. Environmentally friendly measures include the following:

- Use recycled paper, in photocopiers and for forms. The ecologic advantage of recycled paper is the protection of resources and a reduction in waste disposal, saving approximately 90 percent of water consumption and wastewater pollution and 80 percent of the energy demand.

- Restrict use of white (bleached without use of chlorine!) paper.

- Use envelopes, folders, and files made from recycled paper.

The following are responsible uses of paper:

- For in-house communication, use phone and E-mail rather than fax or letter.

- For notes and letters, use both sides of a sheet of paper.

- Use the reverse side of sheets for notes.

- Reduce copying, especially by copying on both sides or scaling down.

- For mailing, use smallest possible envelope (also saves postage).

- Use postcards if possible.

- Do not accept expensive brochures (ask for recycled paper).

- Do not accept unwanted flyers or advertising.

Writing Materials. Give preference to pencil whenever possible. If crayons are used, give preference to products manufactured using food colorants. Felt-tip pens containing solvents are a problem; typically, the casing is made of plastic and they do not rot. Ballpoint pens are not recommended if the casing is made of plastic. In case you must use them, they should be of the refillable type.

Give preference to pencils or crayons as text markers rather than highlighting markers. Use water-based products (without organic solvents) for correction fluid.

Cleaning and Eating. The choice and use of cleaning agents is influenced by the following:

- Reduce number of cleaning agents.

- Give preference to low-waste products.

- Consider central refill stations for cleaning agents (collection and use of residues in containers).

How can waste be reduced in the kitchen? Instead of drinks in cans or nonreturnable bottles, prefer reusable bottles made of glass. Avoid using disposable kitchen utensils (also includes plastic plates).

Responsible Use of Water.
- Use water-saving fittings.

- Reservoir for flush should not exceed six litres.

- Control faucets regularly.

- Dishwashers should be filled completely; use water-saving types.

- Consider using rainwater for toilet flush.

Responsible Use of Energy.
- Room temperature should not be higher than 20°C.

- Reduce central heating (and air conditioning) during the night and on weekends.

- Hot water heating should not be above 45°C.

- Draw blinds overnight (up to 10 percent reduction in energy costs).

- Use daylight (avoid thick curtains).

- Use energy-saving bulbs.

- Consider using solar energy.

- Consider company bicycles for short distances.

- Consider carpooling to work.

Waste Disposal. In principle, waste is not dangerous and presents no special problems in disposal. The exceptions are batteries and expired samples of medicinal products in which special arrangements apply (i.e., disposal through the company itself or through a pharmacy). The quantity of expired samples should be reduced by regular inventorying and by adapted ordering.

Waste mainly consists of

- Household waste (e.g., newspapers, magazines, paper, plastic materials, and glass)

- Packaging materials and cardboard

- Kitchen waste and scraps

Provided that recycling is ensured, collection via separate containers should be available for the following:

- Paper/cardboard: Newspapers, magazines, books, writing paper, brochures, forms, wrapping paper, outer cartons (Paper must be dry and clean, without large labels or foil.)

- Plastic materials (e.g., packaging material like chips and foils)

- Green, brown, and white glass (in separate containers for each type)

- Aluminum

- Polystyrene®, Styrofoam®

- Kitchen scraps (fruit, vegetables, coffee, tea) (These can be either fed to agricultural animals or composted.)

A Special Focus on Regulatory Affairs

A dossier will contain over 100,000 pages, not counting the number of copies required for parallel review by different departments at the Regulatory Bodies. Hopefully, electronic submissions and electronic review tools can help to overcome this situation in the future; in the meantime, the adequate use of paper should be a must for all Regulatory Affairs employees. When planning dossiers/submissions, one should always check whether the extra cover sheet, separation carton, or plastic wallet is really required for the quality of the dossier/submission. This is the responsibility of Regulatory Affairs because there have been cases in which Regulatory Bodies for administrative reasons and/or environmental considerations have renounced on receipt of additional copies or even certain parts of the dossier (e.g., literature). The German Regulatory Body, for example, fought against the flood of plastic wallets for documents. One should not hesitate to argue this point with the Regulatory Bodies. Many companies already successfully combine the cover sheet and separation pages by imprinting the cover sheet information on a different-colored stock that can be inserted/imprinted by the photocopier. This saves also space and freightweight. In the EU, it is perfectly sufficient to use cover sheets/separation pages only for the major parts of the dossier instead of each subsection/binder/document.

Note

1. The general measures suggested in this chapter are based on Daschner, F. 1994. Umweltschutz in Klinik und Praxis. Berlin: Springer-Verlag.

POLICY 15.
THE POLICY ON GLOBAL DOSSIER

Why Is This Policy Needed?

Can you produce the complete, presently valid documentation for all of your company's medicinal products on the spot, let us say, for an inspection by a Regulatory Body? Or, if Marketing decides to register a medicinal product in a new market, is the documentation ready? Or, what if a Regulatory Body decides to call up a whole range of products for immediate renewal (as has happened in the Near East)? Are you sure that you have a good overview of all product changes, including labeling, in order to keep the documentation harmonized during the entire life cycle of the medicinal product? If any of these questions have started you wondering, read on!

The Global Dossier—Can It Be Done? (1)

Definitions

Because there is big intercompany and intracompany variability concerning the use of terms like *submission, standard dossier,* and *global dossier,* definitions will be given first.

- Document for regulatory purposes: Any document that is intended for regulatory purposes (e.g., application for clinical trial authorization or application for marketing authorization).

- Dossier: A compilation of documents relevant for a specific regulatory purpose (e.g., application for clinical trial authorization or application for marketing authorization) in specified country(ies) for a developmental or marketed medicinal product in a structured form (i.e., submission-like). If applicable, it is a subset of the global dossier. The dossier is the basis for the submission(s).

- Global Dossier: A compilation of all documents required for international regulatory purpose(s) for a developmental or already marketed medicinal product. It is maintained continuously throughout the life cycle of the medicinal product and serves as a repository for the generation of dossiers and submissions.

- Submission: A country-specific compilation of documents for a specific regulatory purpose (e.g., application for clinical trial authorization or application for marketing authorization) for a developmental or marketed medicinal product in a structured form according to national regulatory requirements. It is based on the dossier, or, if applicable, the global dossier. It may contain additional national documents (e.g., national leaflets or application forms).

Good Regulatory Practice

Here we will concentrate on one single aspect of GRP, namely the necessity and feasibility of a global dossier, which has been discussed since at least the 1980s (2). What is the regulatory environment as we approach the end of the century?

Harmonization efforts have certainly accomplished a lot, but many areas still need to be harmonized. The regulatory environment outside the EU, Japan, and the United States is evolving quickly and not in a harmonized way, especially in Eastern Europe and the Far East. The concepts of joint review and mutual recognition in the EU have been shown to work in some cases, but they are more often the exception rather than the rule. The industry in general is confronted with a growing flood of regulations; thus, there are calls for deregulation (3). Medical and regulatory terminology is undefined and sometimes misleading; hence, there are initiatives for standardization (4).

Therefore, there is a definite need for a global dossier in order to cut costs, time, and capacity involved on the side of industry and of the Regulatory Bodies. On the other hand, the regulatory environment is not altogether favourable to such a task.

What factors must be taken into account for a global dossier?

- Costs (time, capacity) reduction for the industry and the Regulatory Bodies

- Harmonization efforts ongoing

- Regulatory environment outside the EU, Japan, and the United States evolving unharmonized

- Joint review/mutual recognition not fully implemented

- Deregulation requirements

- Terminology unharmonized

Therefore, the generation of a global dossier should be attempted only under the following conditions: The company's aim should be at least the EU and the U.S. markets (special requirements in Japan must be considered separately); otherwise, the format for the target market should be applied directly. Both international and national regulatory know-how are a prerequisite. The company must arrive at a company position on how to interpret and implement regulatory requirements, also comprising the actual state of the art. Last, but not least, a documentation management system should be in place that enables the modularization and standardization of documents regarding their format and content in order to facilitate the output in submission formats as required. A quality system to ensure quality (and standardization) of work results, in this case, dossiers and submissions, and procedures within Regulatory Affairs is advisable.

The process of developing a global dossier is in itself beneficial. The internal discussion process, which will have to be initiated as a consequence of the decision to produce a global dossier, will result in an increased awareness of the pertinent regulatory requirements. This increased awareness will eventually be reflected in procedures, job descriptions, and the organizational structure of the company. Other benefits may be a reduction in the time to market and a greater degree of harmonization concerning claims in the target countries, which facilitates later maintenance.

The term *global dossier* might signify any of the following approaches: The global dossier might be completely virtual (i.e., evolving during medicinal product development) and existing only as the total sum of documents for regulatory purposes required to fulfill the needs of the EU and the United States, with no need for a specific structure except clear identification of document content and position in the EU or the U.S. dossier. The virtual approach is facilitated by electronic document management and archiving systems with quick output in the required format (e.g., paper, compact disc [CD]), magneto-optical platter [MO] or Computer-Assisted New Drug Application [CANDA]).

The term *global dossier* may also signify a dossier compiled at the end of medicinal product development in a structured format. Based on the output format, the company may choose to produce the specific submission directly or else to supply its subsidiaries with a dossier and leave the fine-tuning to them. In the latter case, the format might be (a) company-specific, (b) EU–like with additions/omissions for the U.S. submission, or (c) U.S.–like with additions/omissions for the EU submission.

This discussion in fact shows that unless global harmonization of requirements is achieved, there is no absolutely true solution to this problem, but it is the only best solution for a specific organization!

The question of a company specific structure versus an EU or U.S. structure may be elaborated on a bit further. Obviously, the EU and U.S. submission structures are easily available (5), whereas the company specific dossier structure must be generated specifically. This process, as well as maintenance, involves certain costs in terms of discussion, consent, and updates. Direct implementation of the EU or U.S. structures and terminology may result in a greater degree of regulator orientation. On the other hand, there will be problems associated with this approach, like resistance ("not invented here"), cultural differences, and company politics, whereas the company solution might be regarded as a fair compromise and, therefore, easier to accept in-house. However, it may be associated with the danger of a lower success rate with the regulators. Apart from the different structure of the dossier, there are some differences between the EU and United States that should be kept in mind (see Table 1).

With a different dossier structure, the TOC and the summary part of the dossier will have to be written specifically for the EU or the United States, and the respective pertinent application forms need to be completed.

Table 1. Comparison of EU and U.S. Structures and the Global Dossier

Structure of Global Dossier	Structure of an EU Submission	Structure of a U.S. Submission
Application Forms	EU specific	U.S. specific
Table of Contents (TOC)	EU specific	U.S. specific
Summary Part	(may be submitted)	required
Expert Reports (evaluation part)	required	(may be submitted)
Written Summaries	(may be submitted)	required
Marketing History	N/A	required
Labeling Text with Annotations	N/A	required
GMP/GLP Statements	part of expert report(s)	separate statements
Statistical Section/Summary	integrated	separate section
Case Report Forms	on request only	required
Environmental Assessment	toxicological part	CMC part
Method Validation Package	integrated	separate package

Some real differences still have not been harmonized because they are based on the approach of the different reviewers. In the EU, the reviewer relies on the evaluation by the company's expert(s) and uses a top-down approach via the expert reports and the tabulated summaries (written summaries if available) when accessing the documentation. The United States, on the contrary, though requiring summaries, relies primarily on the FDA reviewers' evaluation of the data. However, an acceptable solution to this difference is to use the U.S. summaries in the EU dossier in place of written summaries and to include the evaluation of the expert in the U.S. dossier. The marketing history and the annotated labeling text are not required by the EU, as labeling claims are to be justified in the expert report. Including this information in the EU dossier, however, does no harm. Concerning GMP/GLP statements, the statistical section, the methods validation package and, if applicable, the microbiology section, the difference lies not so much in the required data but in the presentation as a separate package or explicit single statement(s) in the United States versus integrated presentation in the EU, or in a different position in the dossier, as in the case of the environmental assessment. Case report forms are definitely considered to be a part of the application basis, but only by the United States. Again, this is a result of the different approach to the review.

Notes

1. Based on Dumitriu, H. 1996. Good Regulatory Practice. *Regulatory Affairs Journal* 7 (10):827–831.

2. Selected articles including the following:

    Cartwright, A.C. and Zahn, M. 1995. The Format and Content of a Global Chemical Pharmaceutical Documentation—A Proposal. *Drug Information Journal* 29:1225–1236.

    Dumitriu, H. 1995. The Industry View of International Standardization of Regulatory Dossiers. *Drug Information Journal* 29:1125–1132.

    Gurien, H. 1991. A Module System for the Preparation of International Dossiers, Manufacturing, and Controls (New Chemical Entities). *Drug Information Journal* 25:285–287.

    Jackson, D.K., Piasecki, S., and Adornato, F.A. 1989. Regulatory Perspective on Worldwide Marketing Authorization Applications. *Drug Information Journal* 23:81–86.

Margerison, R. 1989. Recommendations for a Truly International Registration Dossier. *Drug Information Journal* 23:417–420.

McKenna, K. 1989. An Overview and Comparison of the U.S. and EEC Chemical and Pharmaceutical Requirements for the Marketing Authorization/New Drug Application. *Drug Information Journal* 23:371–377.

McKenna, K. 1989. Working Group 2: The Final Dosage Form—A Model International Registration Dossier. *Drug Information Journal* 23:529–538.

O'Brien, M. 1989. U.S. and EEC Requirements for Documenting the Stability of the Active Constituent. *Drug Information Journal* 23:411–416.

Ostmann, M. 1996. Standardization of Report Formats for Chemistry Pharmacy Documents. *Drug Information Journal* 30:201–206.

Ramsay, A.G. 1989. Working Group 1: The Active Constituent—A Model International Registration Dossier. *Drug Information Journal* 23:515–528.

Schuermans, V., Raoult, A., Moens, M., Heykants, J., Reyntjens, A., Saelens, R., and van Cauteren, H. 1987. International Drug Registration Efforts. *J. Clin. Pharmacol.* 27:253–259.

3.   e.g., by the Center of Medicines Research.

4.   Medical Dictionary for Drug Regulatory Affairs (MEDDRA) initiative to standardize terminology used in drug research.

5.   Notice to Applicants and Code of Federal Regulation.

POLICY 16.
THE POLICY ON IMPORT/EXPORT

Why Is This Policy Needed?

Ever had a study delayed because the study medication could not be imported? Ever had a shortage of product on a major market due to export problems from the country of manufacture? It is necessary to attend to these regulatory details carefully.

Import/Export

National import/export regulations (1) must be adhered to. It should be clarified who is responsible for dealing with import and export licenses. In most internationally operating companies, this task will be assigned to the respective national Regulatory Affairs department.

Note

1. See IFPMA. 1994. *Compendium on Regulation of Pharmaceuticals for Human Use.* Geneva: International Federation of Pharmaceutical Manufacturers Association.

POLICY 17.
THE POLICY ON INFORMATION MANAGEMENT

Why Is This Policy Needed?

Just imagine the damage that could be done to your company due to a leakage of proprietary or confidential information about your company to your competitors. What if the dissemination of false information affects your company's share price? On the contrary, what will happen if reports on, for example, adverse events are not passed on to the responsible person immediately?

There are also the apparently simple things that may happen: Projects are delayed by weeks simply because somebody is on a business trip and nobody else feels entitled (or qualified) to check that person's mail for crucial items. Even if a substitute has been appointed, what is this person to do if nobody is able to grant him or her access to the travelling person's E-mail? Do you think that anybody will acknowledge such an error-prone procedure, if such a gap in the flow of information will cause legal or business problems?

Information Management

Definition

The term *information* in this context means any knowledge on the company's substances or medicinal products that might be relevant for partners within Regulatory Affairs or other contact partners as defined by legal and/or business obligations. This covers information received directly by phone, E-mail, fax, letter, or other route.

The Importance of Information

We live in an information society. The half-life of all scientific knowledge is about five years. Consider that the information exchanged (and the way it is communicated) is what really matters. There is no reality, no truth, except what we say or write about it. Therefore, the basic law for success is to communicate as clearly and honestly as possible.

Some may wish to challenge this statement as the moral climate becomes colder. However, consider what would happen in a world where everybody was lying to everybody else. Clearly, this would be the end of successful relationships, including business relationships. In the long run, the saying holds true: "You can fool all of the people for some of the time and some of the people for all of the time, but never all of the people all of the time". Eventually, there will be severe consequences for your image and your relationships.

Another basic law for success is to inform the relevant people adequately. Now this may seem to be a rather weak statement, especially as the terms *relevant* and *adequately* obviously need to be defined. However, consider that misinformation, typically upper or middle management failing to inform the lower hierarchy, is often at the root of severe mismanagement that endangers the success and survival of companies.

It is often difficult for management to define the borderline between the need-to-know and the nice-to-know. A solution to this problem is not by distributing information but by providing people access to it if they want to (barring of course legal and/or business obligations, secrecy agreements, or very sensitive information). Information must be made accessible but also must be accessed. In this respect, Intranet applications or electronic tools offering discussion groups might be helpful.

Depending on the type of information, a multitude of legal and/or business obligations may apply. Meet these legal and/or business obligations. A careful assessment of such obligations is needed, and procedures should be in place that ensure adequate information handling. One of the most critical aspects in this context is to protect individual rights and personal data. During the medicinal product development process, lots of personal data are being handled, for example, data on the age, gender, and illnesses of patients; therefore, the proper handling of such data is a must.

General Guidelines on Information Management

The management of information by Regulatory Affairs concerning the company's substances and medicinal products must

- Ensure compliance with legal and/or business obligations by the adequate transfer of information to the appropriate contact partners.

- Avoid delays in information, misrepresentation, or lack of information.

- Assure correct interaction with other procedures/ departments involved within Regulatory Affairs and its contact partners worldwide.

Therefore, the following should be adhered to:

- Information on the company's substances or medicinal products should be distributed only on a need-to-know basis.

- With regard to the management of information, the levels of detail and time frames specified by applicable laws and regulations and/or business obligations must be appropriately met.

- All information that may substantially impact the legal and/or business obligations of the company must be communicated in an adequate fashion without undue delay.

- All vital information must be identified and appropriately communicated. A lack of communication of information that could be of importance to the company must be avoided.

- All information communicated by Regulatory Affairs should be true, plausible, or received from a reliable source; however, applicable laws and regulations on the management of information have precedence. If unchecked, incomplete, or dubious information must be passed on, this must be made clear to the recipient of such information.

- The needs of internal and external customers in terms of level of detail and access time must be appropriately met.

- The general business rules concerning secrecy agreements, rules for noncompany personnel, and rules for correspondence and signatures must be appropriately met; for example,

  - Corporate design and forms for correspondence must be adhered to, where applicable.

  - External correspondence should be on standardized paper/format (e.g., company name, logo) and with signatures as required by legal and/or business obligations. Authorizations to sign must be adhered to.

  - Name, function, department, and address of each author must be clearly stated.

  - Electronic mail (telex, fax, E-mail): The rules for external correspondence apply unless the attached documents are signed accordingly and the electronic mail serves only as a cover letter. Texts sent via computer must be signed in the original and stored by the sender. The document must state "signed". For external correspondence, an explanatory sentence must be added: "This text has been created electronically and therefore bears no signature".

  - Going through the mail: A Regulatory Affairs employee should be appointed to open external and internal mail, except documents marked as "private", "confidential", or "personnel matter". In these cases, the department head will make sure proper handling occurs during the absence of the recipient. Mail that does not bear a sufficiently clear address will be opened to identify the recipient by the contents of the document.

  - According to legal and/or business obligations, employees are bound to secrecy toward unauthorized persons concerning information received on the company, subsidiaries, customers, and substances/medicinal products. They must not use such knowledge outside their function for Regulatory Affairs.

  - Generally confidential documents should be marked "for internal use only" and treated accordingly.

- Highly confidential documents should be marked "confidential" and treated accordingly (i.e., stored with limited access, sent only in closed envelopes, mailed as registered letters marked "private"). The envelope must not be marked as "confidential". The copying/distribution of such documents should be by only the permission of the sender.

- The disclosure of confidential information should be made only for regulatory affairs purposes. In all other cases, the relevant departments (e.g., legal) should be contacted to obtain advice.

- The transportation of confidential business papers should be restricted to specifically authorized Regulatory Affairs personnel. Authorizations should be as needed.

- The distribution of circular letters and minutes of regularly recurring meetings should be clearly defined, indicating the record type with a definition, initiator, and a distribution list. Distribution should be on a need-to-know basis.

- Lectures and publications: Special release procedures should be in place for the release of publications. The following items should be checked:

  - Are research results included?

  - Could competitors significantly benefit from the publication?

  - Is release by other concerned departments/disciplines needed?

- Information on suspected or proven risk associated with the company's substances or medicinal products (or information on other substances or medicinal products that could also apply to the company's substances or medicinal products) should be passed on without undue delay to the appropriate department (e.g., Drug Safety). This applies regardless of whether the information refers to approved or unapproved use.

- Insider knowledge: Discrete handling of information that could impact the share price is a must. Also business and/or legal obligations must be appropriately met (e.g., in some countries, it must be assured that important information that could impact the quotation of shares is communicated in an adequate and timely fashion).

- Contact with the Regulatory Bodies (see policy 08)

- Suggestions for improvement: Creativity is a key factor to success and crucial to the continuous process of quality improvement. In order to capture as many ideas as possible, all Regulatory Affairs employees should be encouraged to make suggestions for improvement. Areas for improvement might include profit, cost-effectiveness, ecology, medicinal product development, safety, organization, and procedures or working conditions. Generally, the company should establish procedures to evaluate and decide on the implementation of suggestions for improvement. For suggestions that are approved and implemented, there should be adequate rewards offered in relation to the degree of improvement/benefit to the company.

What Information Is Required for Regulatory Affairs?

- Regulatory Affairs employees should receive the information they require for their everyday work in an adequate level of detail and in a timely fashion.

- There should be clear job titles or functions, with job descriptions, stating responsibilities, the level of authority, and the authority to sign.

- Regulatory Affairs employees should be given the opportunity to take courses and/or attend congresses for their continuing education on a need-to-know-basis (see also policy 12).

- Subscriptions to newspapers and journals should be on a need-to-know basis. Multiple subscriptions to the same publication should be avoided in order to save costs.

- When doing literature queries and requesting copies, rational use should be made of external services; this also applies to the ordering of information on CD-ROM.

- Important visitors, such as those from Regulatory Bodies, to Regulatory Affairs should be announced in advance with the agenda, in order to allow other interested persons to arrange for additional meetings.

POLICY 18.
THE POLICY ON INFORMATION TECHNOLOGY

Why Is This Policy Needed?

Ever had a data security problem? Would it be possible to identify the culprit or are passwords passed around freely in your organization? Are your databases protected from viruses? What would happen if computer hackers or thieves gained access to your most vulnerable databases? How sure are you that nobody is tampering with personal data (e.g., yours)? What are you doing to protect your data (and possibly your job security)?

Information Technology

Regulatory Affairs work is increasingly dependent on and influenced by the use of EDP and/or telecommunication. Most of these electronic tools are beneficial to work as they speed up communication and reduce paper handling. However, it must be ensured that general business obligations and the confidentiality of data are still adhered to. Another concern is the protection of personal data. Procedures in Regulatory Affairs often reflect the paper-based process. It should be considered whether the use of information technology requires new procedures that allow for the parallel review of electronic documents and electronic, instead of paper, archiving. It must be kept in mind, however, that these electronic tools can be used efficiently only with regular training.

POLICY 19.
THE POLICY ON INSPECTION

Why Is This Policy Needed?

Is your company in a state of turmoil every time an inspection is announced? Is there a general feeling that you would rather not be inspected? Then things are not as they should be in your company! Remember, if an inspection identifies serious problems, the Regulatory Body may shut down the plant, recall your products, put you on the black list, and so on.

Inspection

When an inspection by a Regulatory Body takes place, few companies really welcome it. Typically, there will be a certain amount of tension, if not of hasty preparation. However, if the company has an adequate quality system in place and working, an inspection should really not be that different from an audit or a self-inspection—it is a chance for self-assessment and development.

However, there are some simple guidelines that should be adhered to because a negative inspection report can result in severe damage to the company's image and necessitate the closing down of a site or discontinuing production. Like with all Regulatory Body contacts it is important to understand the needs of the representatives of the Regulatory Body—the inspector(s). The Regulatory Body basically requires assurance that the documentation submitted depicts the actual status, that there are no variations to procedures or products that have not been previously notified/approved, and that certain standards are adhered to (e.g., Good Manufacturing Practice [GMP]). The individual inspector will not just take your word for it; it must be ascertained by himself or herself. Typically, limited time will be available in which to carry out the inspection. Thus, good planning is important. There have been cases where the inspector arrived only to find that the manufacture of this specific medicinal product had been discontinued because of renovation or cleaning. With good preparation, this should not happen. Also, the inspector will require your help in showing him or her what he or she wants to see. It is advisable to put together a team of company representatives to accompany the inspector, coordinated by Regulatory Affairs. Regulatory Affairs should write a contact report (see policy 07) in addition to the inspection report in order to record the company's perception of the inspection.

Some general considerations on quality inspections are as follows (1):

- There should be an inspection program/procedure within the Regulatory Body to ensure consistency in inspections and the sharing of expertise.

- If possible, a single, comprehensive inspection should cover all relevant medicinal products of the company.

- Inspectors should be knowledgeable about and familiar with the subject.

- Before the inspection, detailed background information (including product changes, compliance issues) for each company to be inspected should be obtained (e.g., from files and/or databases) and documented in an inspection protocol.

- An inspection plan should be developed.

- Early contact should be made with the company, identifying sites/products/procedures to be inspected.

- Observations should be documented in written form immediately and made known to the company representatives, such as at the end of each day, to allow for the correction of misunderstandings (company representatives should also take notes).

- A written inspection report should be prepared, containing, if applicable, corrective measures already taken by the company, and they should be distributed appropriately.

- Measures taken by the Regulatory Body based on inspection reports must be consistent concerning the extent of such measures.

Note

1. Sanborn, E. 1996. Biological Inspection Program Undergoing Numerous Changes. *Regulatory Affairs Focus* 1 (8):12.

Selected Reading

Jensen, K.B. 1995. Good Manufacturing Practice Inspection in Europe, in Light of the New Central Agency and Current International Agreements. *Drug Information Journal* 29:1211–1216.

POLICY 20.
THE POLICY ON LABELING

Why Is This Policy Needed?

How sure are you that the labeling of your medicinal products in all markets reflects your company's actual state of knowledge? Are you open to liability claims or pharmaceutical critics? Would you feel embarrassed if suddenly confronted with your company's patient leaflets used in developing countries? How certain are you that texts in foreign languages adequately inform patients?

Labeling—Some Considerations

The primary customer for labeling is the patient and/or the professional. The person responsible for placing the medicinal product on the market must ensure adequate information to the best actual state of knowledge. Information on the medicinal product is required for correct use and understanding the possible risks associated with it, but also, if applicable, with noncompliance or abuse.

Not all countries require patient and/or professional information, but the number is increasing. Therefore, the Regulatory Affairs professional is confronted with the task of

- Putting the actual knowledge concerning the product in a nutshell

- Stating the same information twice: in scientific terminology for the professional and in normal, everyday language for the patient

- Informing about possible risks without encouraging noncompliance

Summarizing the Actual State of Knowledge

Summarizing the actual state of knowledge should present no problem if the medicinal product has really been proven, through well-controlled studies, to be effective in the claimed indications. It may be difficult if the claims are too flowery. However, there are cases, such as antibiotics, when the simple listing of susceptible germs will elongate the text considerably. With the amount of information required today, a very long text may present some technical difficulties, as it must fit in the package without unduly increasing the size of the package. However, there are natural limits by what will fit into a pharmacist's storage drawers. Using smaller print may be a solution, but there are limits set by regulations. Regulations also cover the information to be presented. Items to be covered by labeling include the following:

- Pharmaceutical and therapeutic category
- Prescription status
- Active ingredient(s)
- Other excipients with pharmacodynamic or medicinal significance
- Other excipients
- Storage and stability
- Storage and stability after opening the container
- Information on handling, preparation of final dosage form, opening, and measurement of doses
- Mode of action
- Toxicological information
- Pharmacodynamic properties
- Pharmacokinetic properties
- Interaction with other medicinal products and other forms of interaction (e.g., caffeine, alcohol, nicotine, and food)
- Indications
- Posology and method of administration
- Contraindications
- Special warnings and precautions for use
- Undesirable effects
- Use in women with childbearing potential; use during pregnancy and lactation
- Tolerance

- Dependence
- Off-label use
- Overdose
- Interference with laboratory tests
- Warnings

Identical Information for the Patient and the Professional

Usually, it is not the professional information that presents problems. However, it really becomes difficult when you get to the patient leaflet and try to explain the same things as in the professional information (e.g., the mode of action of calcium antagonists) while also trying to avoid scientific terminology. Studies conducted on terminology indicate that the majority of patients do not fully understand the information presented in patient leaflets, so this remains a challenge.

Information on Risks Without Encouraging Noncompliance

Today, Regulatory Bodies require thorough information on contraindications, side effects, and interactions. Rare and very rare effects and even suspected effects must be mentioned—nearly all medicinal products will have an impressive amount of this negative information associated with them. Where the professional will be able to appreciate the subtle differences between rare and very rare side effects, the result may well be noncompliance from the patient. However, adequate information must have precedence because of possible liability issues. Harmonization within a group of medicinal products with the same active ingredient or belonging to the same group should be attempted.

Additionally, for companies operating in more than one country, the Regulatory Affairs professional is responsible for the harmonization of the labeling. This means that, barring active resistance by the Regulatory Body, all national labeling for a medicinal product should contain the same essential statements. This is required not only by ethics, but is also necessary to guard against liability suits and criticism by pharmaceutical critics.

Last, but not least, advertising/promotional material must comply with labeling (see policy 24).

Selected Reading

Advertising and Promotion. 1995. *Regulatory Affairs Journal* 6 (4):342.

AMELIAs Available. 1995. *Regulatory Affairs Journal* 6 (4):335.

Amery, W.K. 1995. Van Winkel M: Patient Package Inserts for Prescription Drugs in an International Pharmaceutical Company. *Drug Information Journal* 29:51–60.

Kendall, V. 1995. Medicines and Patient Information. *EPLC Pharma Law Report* No. 15.

Labeling of POMs. 1995. *Regulatory Affairs Journal* 6 (3):277–280.

Reijnders, P. 1995. Packaging and Labeling: Both Distributor and Marketing Authorisation Holder May Be Named. *Regulatory Affairs Journal* 6 (7):591–592.

POLICY 21.
THE POLICY ON OUTSOURCING

Why Is This Policy Needed?

In case of a crisis, would you know to which CRO to turn? Are you sure that you have a good knowledge of the CROs available for regulatory affairs work, the quality, cost-effectiveness, and reliability of their work?

Outsourcing

It is probable that at some point in time regulatory affairs work will have to be outsourced. *Outsourcing* in this context means insourcing, co-sourcing, or outsourcing of Regulatory Affairs work to external parties. Possible reasons might include, for example, a shortage in personnel, a lack of a specific expertise, or crises hindering Regulatory Affairs from providing the necessary functions as usual. The quality system should provide procedures in this case, otherwise there would be a danger of circumventing the quality system by temporary personnel or CROs.

It is the responsibility of the outsourcing person or department to ensure this by adequately informing the consultant or CRO staff about the quality system. Consider providing special training. It is advisable to prepare a standard contract that binds the consultant or CRO to all applicable business and legal obligations, as well as the standards set out in the quality system on behalf of the outsourcing person or department.

When considering outsourcing, bear in mind that you pay others to broaden their experience at your expense. On the other hand, by using a CRO, you may benefit from the experience of others. In all cases, enough expertise must be developed and maintained in Regulatory Affairs to evaluate and monitor CRO performance.

POLICY 22.
THE POLICY ON PERIODIC SAFETY UPDATE REPORT

Why Is This Policy Needed?

Once a medicinal product is marketed, do you have available the document that is a prerequisite for renewal of the marketing authorization?

Periodic Safety Update Reports

Apart from regular reporting on adverse reactions, regular safety updates are requested by both the EU and the United States. From a regulatory affairs point of view, this is yet another document for regulatory purposes, for which content and format should be defined by in-house standards. Typically, the documents will be generated under the joint responsibility of Drug Safety and Regulatory Affairs. For the sake of harmonization and in order to save time and effort, attempt to create one such document for each medicinal product (instead of one for each Regulatory Body) and update it regularly.

In all cases, Regulatory Affairs should closely monitor the possible impact of the information contained in the safety update on the labeling as part of the dossier, if applicable, the global dossier, the submissions, and already existing marketing authorizations.

POLICY 23.
THE POLICY ON PROJECT ASSIGNMENTS

Why Is This Policy Needed?

Do you frequently receive phone calls from people looking for the responsible Regulatory Affairs manager for a specific medicinal product? Are responsibilities within Regulatory Affairs clear to all Regulatory Affairs employees and also to the scientific disciplines, departments, affiliates, and licensees with whom Regulatory Affairs works? Think of all the time and energy spent when a simple listing could work wonders in this respect.

Responsibility

Quality is also about responsibility; therefore, clear project assignments are a must, as well as clear indications of alternates in case of illness or vacation. For major Regulatory Affairs functions, this may be transparent from the organizational structure. A list of responsible Regulatory Affairs managers per developmental and/or marketed substance/medicinal product should be made available to each Regulatory Affairs employee, as well as to frequent contact partners, especially scientific disciplines, departments, and affiliates. Think of all the time that can be saved if these people can easily identify the correct contact partner within Regulatory Affairs. Electronic tools may be helpful to not only generate and maintain the list, but also to make queries for different types of information possible and to allow access by and/or distribution to interested parties.

POLICY 24.
THE POLICY ON PROMOTION/
ADVERTISING COMPLIANCE

Why Is This Policy Needed?

Can you swear that all promotion/advertising for your company's medicinal products is in compliance with registered labeling and local legal requirements? If not, you are open to liability suits. It is not your responsibility? You had better make it your business or at least get involved to keep out of danger.

Promotion/Advertising

Compliance of promotional/advertising material with registered labeling is not always considered to be the responsibility of Regulatory Affairs. However, bearing in mind that liability suits, especially for over-the-counter (OTC) products, might be based on promotional/advertising material, regardless of patient information supplied with the medicinal product, a company would be well advised to ensure that the promotional/advertising material is harmonized and complies with registered labeling.

Promotion/advertising is restricted by regulations for narcotics and/or prescription only medicinal (POM) products in order to discourage the use of products that might lead to tolerance and/or abuse or increase pressure on budgets for reimbursement. However, also with regard to OTC products, guidelines must be observed, whether issued by Regulatory Bodies or through self-control by industry associations. Generally speaking, promotional/advertising material must not be misleading and must not arouse fear of illness. Also, direct comparison to competitors is usually not allowed. Care should be taken that, together with the promotional/advertising material, the minimum required information on safety and risks is given. The quantity of print or—for commercials in radio/television—the time allocated to this statement should be adequate.

Selected Reading

Advertising and Promotion. 1995. *Regulatory Affairs Journal* 6 (4):342.

Kamp, J.F. 1995. An Advertising Agency Perspective of Food and Drug Administration Pharmaceutical Regulation. *Drug Information Journal* 29:1301–1306.

POLICY 25.
THE POLICY ON REGULATIONS AND GUIDELINES

Why Is This Policy Needed?

How can you expect to develop medicinal products successfully and to get clinical trial license and marketing authorization when you do not closely monitor the actual regulatory environment? Do not just take anything the Regulatory Bodies throw at you lying down—get involved; they expect and invite discussion. After all, you are the people who really are knowledgeable in the company's medicinal products.

Regulations and Guidelines

The pharmaceutical industry is heavily regulated: Placing medicinal products on the market and conducting clinical trials in humans without marketing authorization and clinical trial license is prohibited by law. The flood of new drafts/final documents issued by the Regulatory Bodies to specify the conditions for marketing and clinical trials and to provide guidance to applicants is ever rising.

The regulatory environment must be carefully screened for applicable national (or for the European Union [EU]) regulations. The guidelines of the World Health Organization (WHO) and International Conference on Harmonisation (ICH) apply. It is important to realize that not only final documents with legal status (e.g., laws and regulations) must be adhered to; documents that still have to be transferred into national law (e.g., EU Directives and ICH guidelines) must also be adhered to. The same applies to guidance documents—Guidelines, recommendations, and Points to Consider documents—even if only drafts, as they define the actual state of the art: Regulatory Bodies will use them as a basis for decision making. In a court of law (e.g., in liability suits), the fact of not considering such final or draft documents may considerably weaken your position.

It is the job of Regulatory Affairs to monitor closely and evaluate the regulatory environment. In order to make sure that you obtain all required information on a regular basis, it is suggested to combine external services as needed. Unfortunately, no single vendor covers all Guidelines and Regulations worldwide. Another reason to use parallel sources is that the quality or completeness of documents received may be an issue. Up to now, Guidelines from Regulatory Bodies are received in paper format. However, U.S. and EU documents may also be obtained in electronic format via the Internet.

For internationally operating, large organizations, it is advisable to maintain a repository of the relevant Guidelines in a database with bibliographical information and full texts. Full electronic texts are a must, as experience shows that keywords are not sufficient for all queries. Also, the process of assigning adequate keywords to each document and the maintenance of the keywords list or thesaurus is very capacity intensive (more than one man/year), not to mention intra- and interperson variability in assigning keywords, which could impact the quality of query results.

Regulations/Guidelines may be obtained from the following addresses (this is not a complete list; for national regulations, contact the national Regulatory Body or national industry association):

AAC Catalog of Regulatory Documents
   AAC Information Services Division
   7575 Wisconsin Avenue
   Suite 850
   Bethesda, MD 20814
   Tel: +1 301 986 4440
   Fax: +1 301 986 4448

Celex/Eurolex
   Amt für Amtliche Veroeffentlichungen der Europaeischen
   Gemeinschaften
   Dienststelle Vertrieb
   L-2985 Luxembourg

Drug Information Association
   P.O. Box 3113
   Maple Glen, PA 19002
   Tel: +1 215 628 2288
   Fax: +1 215 641 1229

EuroDirect Publications Office
   Room 1205 Market Towers (Information Centre)
   Medicines Control Agency
   1, Nine Elms Lane
   London SW8 5NQ
   Tel: +44 71 273 0343
   Fax: +44 71 273 0334

European Federation of Pharmaceutical Industries'
   Associations (EFPIA)
   Avenue Louise 250 Boite 91
   1050 Brussels, Belgium
   Tel: +32 2 626 2555
   Fax: +32 2 626 2566

European Pharma Law Centre LTD (EPLC)
18-20 Hill Rise
Richmond, Surrey TW 10 6 UA
Tel: +44 81 948 3262
Fax: +44 81 332 8992

Information Medicales & Statistiques
IDRAC Division
Immeuble La Défense-Bergères
345 Avenue Georges Clémenceau
TSA 30001-9282
Nanterre CTC Cedex 9
Tel: 1 41 35 10 00
Fax: 1 41 35 13 44

Interpharm Press
1358 Busch Parkway
Buffalo Grove, IL 60089

Regulatory Affairs Professionals Society
12300 Twinbrook Parkway Suite 630
Rockville, MD 20852
Tel: +1 301 770 2929
Fax: +1 301 770 2949

Rostrum Regulatory Guideline Service MCRC Group Ltd.
1, Mildmay Road
Romford, Essex RM7 7DA
Tel: +44 708 735191

Selected Reading

Nightingale, S.L. 1995. Challenges in International Harmonization. *Drug Information Journal* 29:1–10.

Versteegh, L.R. 1995. Bringing Information on Evolving FDA Standards to Bear on Ongoing Drug Development Programs. *Drug Information Journal* 29:1097–1104.

Worden, D.E. 1995. The Drive Toward Regulatory Harmonization: What is Harmonization and How Will it Impact the Global Development of New Drugs. *Drug Information Journal* 29:1663S–1679S.

POLICY 26.
THE POLICY ON REGULATORY STRATEGY

Why Is This Policy Needed?

Are you sure you have understood your project team's targets? Is it to register as soon as possible, even if some claims have to be dropped, or is it the exact wording of the labeling? Is there a major market that has top priority, or is there a strategic need to access all markets at the same time? Are all countries absolutely necessary, even if it means waiting for some trials against yet another gold standard? Is the situation harmonized enough (product status and therapy) to allow an EU procedure, or should you go national?

Regulatory Strategy

Determining the regulatory strategy is one of the major functions of Regulatory Affairs. This function offers high visibility to the Regulatory Affairs professional in the organization.

The ultimate goal is to receive qualified approval quickly by meeting only the necessary requirements. Regulatory Affairs should always consider the company's need to save time and costs. A certain risk may be acceptable if it is stated clearly to all concerned parties (e.g., project team and upper management).

The quality of a regulatory strategy is defined by timely approvals of target claims in target countries. Therefore, it is necessary to clarify in advance with upper management and marketing, for example, what the major targets really are.

Before generating a regulatory strategy, clarity must be reached regarding the following items:

- Target SMPC
- Target countries
- Applicant(s)
- License holder(s)
- Trademark(s)
- Dosage form(s)
- Strength(s)
- Primary packaging(s)

Ideally during medicinal product development, a target SMPC should be generated and maintained, and agreement should have been reached regarding the importance of core statements versus exact wording, and the priority of the indications. This will be an important basis in evaluating the chances and, if applicable, the different procedures for obtaining the desired approval(s).

By identifying the target countries, the territory in which the regulatory environment must be screened with regard to therapeutic culture, standard therapy, competitor products, requirements for national trials, or trials with special ethnic groups can be established.

Applicant(s) should be clarified for administrative reasons. The conditions for the possible transfer of a marketing authorization should be assessed. The legal/regulatory obligations of the license holder should be carefully monitored so that licensing does not violate these obligations.

The advance registration of a trademark(s) is desirable, as this may take considerable time. It must be ascertained whether the trademark is not too similar to another, already registered trademark. By inference, the trademark must not suggest a very high quality or make false claims. In the EU, a single trademark should be used.

The dosage form(s), strength(s), and/or primary packaging(s) will determine which dossier(s)/submission(s) can be prepared jointly, and, for EU procedures, which submission(s) can be considered as identical. Also, most Regulatory Bodies require information on marketing history and/or submission status of the (identical) medicinal product.

It is not possible to list all of the considerations that are part of the generation of a regulatory strategy. However, as an example, some questions are given below, which might be considered helpful for generating a regulatory strategy in the EU. If any of these questions must be answered negatively, no EU procedure should be attempted.

- Are the indication(s) defined similarly in target Member States?

- Are therapy standards (e.g., gold standards) either similar or has the product been tested against all of the gold standards in the target Member States?

- Is product status the same (e.g., food, cosmetic or medicinal product) in the target Member States?

- Is the quality of the documentation either good or excellent?

- Is access to at least two Member State markets intended?

- Is the product not a homeopathic?

When deciding whether to use the decentralized or the centralized procedure, remember that biotechnology products MUST undergo the centralized procedure. In the case of innovative products, the applicant may decide to use either the decentralized or the centralized procedure. Given below are the conditions for innovative medicinal products eligible for the centralized procedure. (Confirmation by the European Agency for the Evaluation of Medicinal Products on a case-by-case basis is advisable.)

- Medicinal products developed by other biotechnological processes that in the opinion of the Agency constitute a significant innovation

- Medicinal products administered by means of new delivery systems that in the opinion of the Agency constitute a significant innovation

- Medicinal products presented for an entirely new indication that in the opinion of the Agency are of significant therapeutic interest

- Medicinal products based on radioisotopes that in the opinion of the Agency are of significant therapeutic interest

- New medicinal products derived from human blood or human plasma

- Medicinal products whose manufacture employs processes that in the opinion of the Agency demonstrate a significant technical advance, such as two-dimensional electrophoresis under microgravity

- Medicinal products that are intended for administration to humans and that contain a new active substance that, on the date of entry into force of this regulation, were not authorized by any Member State for use in a medicinal product for human use

An extension of the definition of innovative medicinal products has been suggested with a view to giving more products access to the centralized procedure (1).

Generally, EU procedures become more difficult if the product is a combination because of the different perceptions of combination products in the Member States. The same applies if the active constituent is a known active ingredient in some, but not all, Member States. The submission of confidential information (e.g., Drug Master Files [DMFs] by subcontractors) may also present a problem, if it is not available in the required language or not completely identical.

Note

1. Van Essche, R. with the key support of Prof. Benzi, Member of European Parliament (MEP), 1996. *European Development of Innovative Drugs*. Paper presented at the RAPS meeting, Amsterdam, 22 April 1996.

   Extract from Annex 1 on human innovative drugs (evaluation scale):

   1. Drugs that show therapeutic efficacy for a disease or a symptom for which there is no active drug available

   2. Drugs that show therapeutic efficacy for a disease or a symptom for which an effective drug is already available but whose effect is necessary for a subset of the affected population

   3. Drugs that are more effective and/or show less serious adverse effects than the reference drug of an equivalent therapeutic effect

   4. Drugs that may be given to special groups of patients with increased efficacy or reduced toxicity

   5. Drugs that are presented in a form that is more practical and/or convenient for the patient

Selected Reading

Versteegh, L.R. 1995. Bringing Information on Evolving FDA Standards to Bear on Ongoing Drug Development Programs. *Drug Information Journal* 29:1097–1104.

POLICY 27.
THE POLICY ON SUBMISSION

Why Is This Policy Needed?

Can you submit in less than one month from receipt of the dossier/documentation? Is the layout of the submissions to each country similar or does each Regulatory Affairs employee create his or her own style? How do you think that suits regulators—having to adjust to yet another way to present the data? What if key personnel became ill? Would Regulatory Affairs still be able to do the same high quality submissions?

Dossier/Submission

Definitions

Because there is big intercompany and intracompany variability concerning the use of terms like *dossier, submission, standard dossier,* and *global dossier,* definitions will be given first.

- Document for regulatory purposes: Any document that is intended for regulatory purposes (e.g., application for clinical trial authorization or application for marketing authorization).

- Dossier: A compilation of documents relevant for a specific regulatory purpose (e.g., application for clinical trial authorization or application for marketing authorization) in specified country(ies) for a developmental or marketed medicinal product in a structured form (i.e., submission-like). If applicable, it is a subset of the global dossier. The dossier is the basis for the submission(s).

- Global Dossier: A compilation of all documents required for international regulatory purpose(s) for a developmental or already marketed medicinal product. It is maintained continuously throughout the life cycle of the medicinal product and serves as a repository for the generation of dossiers and submissions.

- Submission: A country-specific compilation of documents for a specific regulatory purpose (e.g., application for clinical trial authorization or application for marketing authorization) for a developmental or marketed medicinal product in a structured form according to national regulatory requirements. It is based on the dossier, or, if applicable, the global dossier. It may contain additional national documents (e.g., national leaflets or application forms).

Goal

One of the major functions of Regulatory Affairs is to apply for and to obtain regulatory approvals (e.g., clinical trial authorization, marketing authorization, and renewal of marketing authorization). Hence the importance of dossier/submission generation. Regulatory Affairs has the primary responsibility for the resulting product, namely, the dossier and/or the submission. The ultimate goal is to receive qualified approval quickly by meeting the necessary requirements only.

Quality Before Time Before Costs!

Quality will be achieved by meeting the necessary requirements through the submission of data sufficient to meet regulatory requirements and to provide assurance on quality, efficacy, and safety of the medicinal product. Time to market is of the essence. Therefore, time will be cut as much as possible during the submission and review phase. It may even be necessary to submit additional data just to speed up the procedure even if it is disputable whether they are really needed. Costs are a factor when defining the necessary requirements. However, quality and time must have precedence.

Quality for a dossier/submission is quality in terms of content, format, and timely finalization. Quality in terms of content must be created by early and continuous input from Regulatory Affairs during the development process via the generation of internal company standards for documents for regulatory purposes and adequate procedures. The actual contents of a global dossier/dossier/submission should be handled by standards and adapted to specific medicinal products/regulatory requirements. If required, the same applies for quality in terms of format.

Responsibilities and workflow for the generation of documents for regulatory purposes, dossiers, global dossiers, submissions should be clear in advance. Use procedures set up as local SOPs.

For locally operating companies, there will be no need to create a global dossier. Also, the dossier and submission generation will usually be combined. For internationally operating companies, there may be two opposing interests in the company:

1. To save time in development and later in maintenance by creating, if possible, one single (global) dossier or, if this is not possible, regional dossiers (e.g., a single dossier for the EU).

2. To customize submissions to the local requirements in order to save time for local approval.

These two opposing interests are often mirrored by the existence of a corporate and a local regulatory affairs function. Corporate Regulatory Affairs will be more knowledgeable about the medicinal product and international regulatory requirements and harmonization efforts, while local Regulatory Affairs will be more knowledgeable about specific local requirements. Therefore, it is advisable to have corporate Regulatory Affairs accompany the development process and generate the dossier and to let local Regulatory Affairs generate the specific submission. In this way, it should be possible to reach the best possible compromise.

As the quality requirements for contents are discussed in other policies, the technical and formal aspects of dossier and submission generation will be focussed on here. (As the requirements for dossier and global dossier, only the term *dossier* will be used.)

Today, as much as 30 percent of the capacity of the Regulatory Affairs department is required for the generation of dossiers and/or submissions. The reason for this capacity is, besides a possible lack of standardization, the paper-based approach, which involves a lot of manual work. The use of electronic tools will hopefully improve this situation.

Dossier generation may involve the following steps:

•   Generation of a TOC

•   Generation of cover pages for individual parts

•   Compilation of copies of documents for regulatory purposes according to the TOC

- Pagination*

- Other imprints (e.g., "confidential", date)*

- Cross-referencing from the TOC and the summary part to the documentation*

- Copying from dossier Master File

- Insertion into binders

- Insertion of separation pages

- Labeling of binders

Submission generation may involve the following steps:

- Generation of the TOC

- Generation of documents for specific regulatory purposes (e.g., application forms, national labeling, proof of payment, list of samples) and inclusion in the submission

- Pagination**

- Other imprints (e.g., "confidential", date)**

- Cross-referencing from the TOC and the summary part to the documentation**

- Copying from submission Master File

- Insertion into binders

- Labeling of binders

- Insertion of separation pages

- Production of a cover letter

- Addition of any other material (e.g., samples)

Technical Aspects of Quality

Bear in mind that the Regulatory Bodies will expect generally used terminology and format in the TOC. Any other format will make the dossier check-in procedure and location of information during the review more difficult for the Regulatory Body, thereby

---

*May also be done as part of the submission generation

**May also be done as part of the dossier generation

losing valuable time for clarification. Also, whenever a header or section is not applicable to the specific medicinal product, this should be so stated rather than the header/section being deleted from the TOC. It is also advisable to adopt the numbering system of the Regulatory Bodies as far as possible.

There is no specific requirement for the cover page, but it is useful to use such cover pages for the whole dossier and the main parts, without overdoing it. Cover pages should state the title and number of the section. Some companies also add the name of applicant, the date of application, the name(s) of medicinal product, the dosage form, and the strength. When cover pages are kept as neutral as possible, they can be printed in advance, which greatly facilitates the process and helps to reduce costs. Also, there may be problems encountered with details (e.g., applicants or trade names might be different from country to country). Some companies combine cover pages with separation pages.

The compilation of copies of documents for regulatory purposes is the most time-consuming step when using a paper-based method. The copying process from originals may lead to deficiencies (e.g., missing pages or bad copying quality). This is where electronic tools are a major improvement. Using an optical archive enables the printing of documents in similar quality as the original and directly in the order of the TOC.

The pagination of a paper file can be done by special machinery or directly by specially equipped photocopiers. Ideally, pagination should be done electronically during the compilation process, including other imprints, such as the date. It may be worthwhile to consider paginating each part or even each volume separately instead of the whole dossier to allow parallel work in order to save time.

When using a paper-based method, cross-referencing can be done only after the documentation has been finalized and paginated. When using electronic compilation tools, hyperlinks make it possible to do a lot of work in advance or in parallel, thus cutting the time to submission.

Again, the copying process by itself can lead to deficiencies in the resulting copies of the Master File. At least one copy should be checked in its entirety; in other copies, spot checks should be made.

Good advance planning/organization of material, workflow, and rooms is essential for binder preparation (insertion, separation, and labeling). Considering that a marketing authorization application for an NCE consists of 300–600 volumes per copy, electronic tools make it possible to process the printout as it is printed.

Last-minute changes should be avoided as much as possible, especially because of the major impact on the TOC, the pagination, the cross-referencing, and the resulting loss of time. Here, too, electronic compilation tools can be helpful. When using electronic tools, the finalization of the dossier/submission can be started immediately after the last document has been received at Regulatory Affairs. The time to submit depends mostly on the speed of the printer.

Presently, time estimates for dossier generation range from one to six months and one month or less for the generation of the submission. By using electronic tools and adequately standardized procedures, it should be possible to submit documentation within one month after the receipt of the last document at Regulatory Affairs.

POLICY 28.
THE POLICY ON TERMINOLOGY

Why Is This Policy Needed?

Are key terms used with the same meaning within Regulatory Affairs and the company? Does everybody who needs to understand the abbreviations? Better make sure you use the same lingo!

Terminology—Some Reflections on Nomenclature, Standards, and Other

Language is one of our most important tools in life. We use it to communicate objective information, our emotions, our feelings, and thoughts about the relationship to the other parties and what we would like them to do. Any time we communicate, we do it on these four levels simultaneously. What will be received/understood by the other party depends on (a) what has been transmitted and how, and (b) the disposition of the receiver. Obviously, we can control to some extent what and how we transmit, but the disposition of the receiver is largely beyond our control and our information on it is also limited. This very brief summary of a fascinating field makes clear that effective communication is above all clear communication.

Not only should ideas and concepts be clear, but terms and expressions used by the communicating parties should have the same meaning. This is important in order to understand each other. It is not by chance that the expression "speak the same language" is synonymous with understanding each other. This is also why technical languages and specialist terminologies are developed and shared by various disciplines. Using agreed-on terminology helps to communicate effectively. It also speeds up processes, as misunderstandings or lengthy definition processes are reduced.

Especially for Regulatory Affairs, the importance of language and terminology cannot be overestimated. For example, *drug*, a term commonly used in the United States for a finished product, should not be used in the EU, as it recalls drug abuse and addiction. Instead, the term *medicinal product* should be used. The term *Computer-Assisted New Drug Application (CANDA)* can signify anything from a few texts on a floppy disc to a virtual dossier, including image technology, hyperlinks, coded text, and databases. The terms *dossier, submission,* and *documentation* are often used indiscriminately. These are only a few examples to illustrate the necessity of clear terminology and definitions. Consider also that Regulatory Bodies do not approve your medicinal products (most reviewers will never see them even if samples are required with the submission). They approve THE INFORMATION communicated. Your communication with the Regulatory Body vastly depends on how clearly, interestingly, and convincingly you state your case in written form. Thinking of adverse reaction reporting and labeling claims, the impact of terminology becomes even clearer.

The need for a common terminology has been realized by the Regulatory Bodies. The ICH is pursuing a medical thesaurus for international premarketing and postmarketing regulatory purposes based on the Medical Dictionary for Drug Regulatory Affairs (MEDDRA) developed in the United Kingdom, replacing Coding Symbols for a Thesaurus of Adverse Reaction Terms (COSTART) and the World Health Organization's (WHO's) Adverse Reaction Terminology (WHOART) and including International Classification of Diseases (ICD) 9 and 10. It covers side effects, adverse drug reaction reporting, indications, over-the-counter (OTC) classification, and clinical trial reactions. It will be the basis for a common terminology for Regulatory Bodies and industry in the ICH triad. Implementation will be in phases in the EU, Japan, and the United States. The availability of common terminology will enhance communication and harmonization, and simplify regulatory processes.

Other efforts include a project to develop software for a multi-language version of medical terminology in the regulatory process under the EU on-line system EudraLex. In the United States the National Library of Medicine (NLM) and the Agency for Healthcare Policy and Research (AHCPR) are evaluating with regard to the description of real clinical data NLM's Unified Medical Language System and 38 other vocabularies including Systematized Nomenclature of Medicine (SNOMED) and the Read codes, both structured nomenclatures for medical use. A common language within the company and within the Regulatory Affairs community is of great value to assure common understanding, to speed up processes, and to make joint efforts possible.

Integrating dictionaries may present some problems, as responsibility will typically lie within different departments (1). It must be kept in mind that the development and maintenance of a thesaurus is capacity and time-consuming. Therefore, preference should always be given to thesauri developed and maintained extramurally. As long as several dictionaries are used by the Regulatory Bodies, translation between them remains an issue. Some companies have already established terminology systems of their own (e.g., Merck's autoencoder (2) or use systems that can convert terminology (e.g., Takeda's TRAC (3). Also, non-exhaustive compilations of abbreviations most often employed in drug regulation are offered as reference tools (4).

Notes

1. Jolley, S. 1995. Clinical Safety, Administration and Data Systems: How Should They Be Integrated? *Drug Information Journal* 29: 661–663.

2. Gillum, T.L. 1995. George, R.H, Leitmeyer, J.E. An Autoencoder for Clinical and Regulatory Data Processing. *Drug Information Journal* 29 (1):107–113.

3. Sakurai, Y., Kugai, N., Kawana, T., Fukita, T. and Fukumoto, S. 1995. A Comprehensive Adverse Events Management System for the Pharmaceutical Industry: The Takeda TRAC System. *Drug Information Journal* 29:645–659.

4.   e.g., Carre Llopis, M.C. and Jimenez, V.J. 1995. Abreviaturas, Siglas y Acronimos en el Mundo de los Medicamentos (Abbreviations, Initials, and Acronyms, in the World of Drugs). *Farmacia Clinica* 12:62–75; Fleschar M.H. and Kimbal R.N. 1996. *Glossary of Biotechnology Terms*. Basel, Switzerland: Technomic Publishing AG.

POLICY 29.
THE POLICY ON TOOLS

Why Is This Policy Needed?

Is Regulatory Affairs enthusiastic about new programs and databases, but do they seldom work out as expected? Or are you barely using computers except for word processing? In both cases you want this policy.

Tools for Regulatory Affairs

Our culture deals primarily with information. Also, Regulatory Affairs departments process and add to the information received from the scientific disciplines and the Regulatory Bodies (e.g., creating the dossier structure and cross-references) and generate information (e.g., evaluations of Guidelines).

Today, 60–80 percent of this information is handled in paper format. However, with word processing and E-mail, there is a strong trend toward electronically generated and processed information. We also wish to make the best use of the existing information (e.g., via databases), which allow query and the retrieval of data in varying output formats.

Therefore, the Regulatory Affairs professional will increasingly be confronted with the question of which electronic tools might be beneficial. He or she will be requested to come up with specifications and requirements and later on be expected to learn how to use these tools efficiently.

Apart from the most commonly used areas (i.e., word processing, calculations, and graphical programs), electronic tools might be beneficial in the areas of archiving (e.g., DAMOS, Documentum®, Interleaf®); document management (e.g., Documentum®, Interleaf®, Saros); dossier compilation (e.g., Documentum®, Interleaf®, Pharmbridge®); product information (e.g.,

PILS); and the tracking of documents and dossiers, marketing history and registration status, regulations, and guidelines (e.g., EPLC, IDRAC®). An internationally operating company may even want to create its own programs to capture regulatory know-how (e.g., Merck's regulatory query tracking system) (1).

When considering and evaluating electronic tools, it is advisable to work closely together with Information Technology specialists. When developing programs, it is helpful to form user groups and to set up test plans as well as to do random testing in order to identify flaws. However, generally speaking, it is advisable to use commercially available tools rather than adapt or develop your own.

The Regulatory Affairs department should have a clear concept of where electronic tools might be most beneficial for this specific department and where the concept fits in with the overall EDP concept of the company. Also consider combining several programs or databases with interfaces in order to avoid the duplication of work (e.g., for data entry).

Note

1. Michalak, R.A., Mathews, R.A., Leslie, W.D., Jenkins, A.A., Mehta, M.S., and Asbjorn, D.T. 1996. The Query Response System: A Database Which Captures Regulatory Questions/Responses. *Drug Information Journal* 30:207–215.

6

Policies

This chapter provides policies that define the basic principles under which Regulatory Affairs is to operate worldwide. Policies reflect the agreed-on quality requirements. They are concise and contain generalized wording. Policies may reference recommendations; guidelines; regulations; and scientific, legal, ethical, or other standards.

Key policies on "policy" and auditing are presented first in order to give the general outline of the quality system. The other topics are presented in alphabetical order. Consideration was given to clustering topics that are closely linked and sometimes overlapping (e.g., documents for regulatory purposes, dossier, electronic dossier, global dossier, and submission; electronic submission, information technology, and tools); however, this was not done because if would have caused too much repetition.

For background information on the topics of policies, see chapter 5, "The Philosophy Behind the Policies". For definitions of items that must be the same throughout the organization, see chapter 7, "Standards".

The reader is invited to adapt the policies and/or standards to his or her organization and function and, if required, to develop quality assurance processes. These processes should form the basis of Standard Operating Procedures (SOPs). If required, further policies may be established; however, they must not deviate from existing policies.

A Policy of Department XYZ

01: Policy

Document Type:	Policy
Document Code:	01
	(enter company-specific code)
Title:	Policy
Date/Revision No.:	DD/MM/YY number xy
Scope:	Global
References:	(enter policies, standards, SOPs of your department/company, or other documents [e.g., guidelines] that should be considered in this context)
	1. Standard for Policy (S-01.01)
	2. Policy on Auditing and Compliance (P-02)
Authorization:	
	Signature of authorized person(s)
	Name of authorized person(s)
	Job title/Function of authorized person(s)
Issue Date:	DD/MM/YY
Implementation Date:	DD/MM/YY

1.   PURPOSE

This policy is a set of rules developed to govern the scope, content, format, and intent of Regulatory Affairs policies, standards, and standard operating procedures (SOPs) as elements of the quality system within Regulatory Affairs worldwide.

2.   DEFINITIONS

The key terms pertaining to this policy should be defined here. As there are no uniform and globally accepted definitions available, please develop your own definitions. In this way the language of the staff of your organization can be incorporated.

- *Policies* define the basic principles under which Regulatory Affairs is to operate worldwide.

- *Standards* are definitions of items that are required to be the same throughout the organization.

- *Standard Operating Procedures (SOPs)* define how policies are implemented and standards are met in daily operations.

3.   STATEMENT OF POLICY

This section covers the actual rulings that should be complied with when working according to the principles of the quality system. For the topic *Policy*, rulings should be available for the following items:

3.1.   Policies will be established as needed to govern the work of Regulatory Affairs worldwide.

3.2.   Policies reflect the agreed-on quality requirements. They are concise and contain generalized wording. Policies may reference recommendations; guidelines; regulations; and scientific, legal, ethical, or other standards.

3.3.   The format of policies is set out by a Standard (1).

3.4.   Standards and SOPs will be established (if required separately, also on a national level) to assure that all Policies are fully implemented and that all Standards are met in daily operations. SOPs must fully reflect the Policies as far as applicable; however, SOPs may surpass them or be more stringent or elaborate, if required. If required, additional Standards and/or SOPs may be created; however, they must not deviate from existing Policies or Standards.

3.5. Policies, Standards, and SOPs are authorized by authorized person(s).

3.6. Audit procedures and adequate feedback mechanisms will be established to assure adequate maintenance of the quality system (2).

4. RESPONSIBILITY FOR POLICY IMPLEMENTATION: NAME AND TITLE/FUNCTION OF AUTHORIZED PERSON(S)

5. RULING PERTINENT TO EXISTING PROCEDURES

6. RULING APPLICABLE IN THE CASE OF OUTSOURCING

7. POINTS TO CONSIDER DURING THE DEPLOYMENT OF THIS POLICY

General Principles for Applying Policies

Before implementation, read the policy and the points to consider carefully. The latter are designed to help you fully understand the meaning of the policy and the reasons why this specific wording has been chosen.

In all probability, it will be necessary to adapt the policy to your specific department/organization. Bear in mind that changes should be carried through by an in-depth discussion process between all interested parties with mutual agreement only. This process is required to ensure a common understanding and interpretation of the policy as well as full commitment for complying with its provisions. Therefore, it is suggested that you document the reasons for changes in your own version of points to consider as a justification document for future reference.

Every policy consists of a cover sheet stating the Document Type, Code, Title, Date, Revision, Scope, References, Authorization, and the body of the policy consisting of

1. Purpose

2. Definitions

3. Statement of policy

4. Responsibility for policy implementation

5. Ruling pertinent to existing procedures

6.   Ruling applicable in the case of outsourcing

7.   Points to consider during the deployment of the policy

The cover sheet fulfills organizational needs by stating the type, code, date, and revision number (= version) of document together with its organizational number (and applicable references) so that exact identification is possible (very important for updates). It also gives the authority of the document and its scope. An issue date (= signature date) is given, as well as the date by when the policy must be implemented.

The purpose enables the reader to understand scope and intent of the policy. Then definitions that are important for understanding the policy are given. A statement of policy that sets the quality standard for this topic follows the definitions. The function(s) having responsibility for implementation are then given, as well as rulings pertinent to other already existing procedures and rulings applicable in the case of outsourcing.

Each policy should be a separate file/document paginated consecutively (suggested format: page x of xx) to assure consistent handling. It is recommended to indicate the author, file name, date, and word processing program in the footer and to indicate the document type (here: Regulatory Affairs Policy), code, date (of revision, not necessarily identical with that of file), and revision number in the header of all pages of the policy.

Cover Sheet

- State the issuing department.

- Document type: "Policy" indicates that it is a policy; therefore, it is a more general type of document than, for example, Standard or SOP.

- Document code: For example, P-21. P signifies Policy, S signifies Standard, and SOP signifies Standard Operating Procedure. The numbers should be given consecutively to ensure correct identification (in conjunction with the revision number). You may wish to consider another numbering system, for example, forming clusters of related topics. However, the adequate handling of such a numbering system becomes difficult over time. As the number of required policies for a department is quite low (because of the general character of policies), a consecutive numbering system is recommended.

- Title: The title should be as short as possible, stating the subject only.

- Date/Revision No.: The date is the date of finalization; the revision is the number of the version. Both help to identify any specific document correctly. It is suggested to count the draft versions only until a final version is agreed. This is then revision 1. Otherwise, the numbers get too high. Once the final version is available, there should be little interest in previous draft versions. However, all signed revisions must be stored in order to maintain and document the quality system. Any modification to a document requires a new revision number with a full printout. It is recommended to exchange the complete document in the Quality Manual rather than single pages. (See also above on recommendations for header and footer of the document.)

- Scope: Defines the range of applicability for the specific organization (e.g., global for internationally operating companies); adapt as required. In the policies set out here, "global" is entered as they contain sufficiently generalized wording.

- References: If other documents, standards, or regulations are referenced, it is part of good practice to clearly identify such documents. For the purpose of easy update, the references are given on the cover sheet and numbered consecutively as they appear in the text.

- Authorization: All functions that are responsible for assuring the standard of quality set out by the policy must authorize the policy by signature to ensure common understanding, mutual agreement, and full commitment. When presented together in a manual, it may seem redundant to sign individual policies; however, for the sake of greater awareness and later maintenance of revisions, it is advisable to handle each policy individually. For each policy, consider whether co-signatures from other function(s) are required. If so, let them participate in the discussion process as early as possible.

- Issue date: The signature date is entered.

- Implementation date: The date when the policy must be implemented; it may or may not be identical to the issue date.

Policy

7.1. Purpose

This policy, together with the policy on auditing and compliance, forms the backbone of the quality system. It defines the structure of the system and its key elements. The opening sentence must be standardized for the sake of harmonization: "This policy is a set of rules developed to . . . " At the end of this section for global policies, the scope is addressed by stating that the policy is applicable for Regulatory Affairs worldwide.

7.2. Definitions

The key terms pertaining to this policy should be defined. As there are no uniform and globally accepted definitions available, please develop your own definitions. In this way, the language of the staff of your organization can be incorporated. The key terms required for a full understanding of this policy are *policies, standards,* and *standard operating procedures.*

7.3. Statement of Policy

This section covers the actual rulings that should be complied with when working according to the principles of the quality system. For the topic *Policy,* rulings should be available for the following items:

7.3.1. Rather than develop a manual governing each and every aspect of Regulatory Affairs work that could possibly impact quality (e.g., length of breaks), another approach has been chosen, namely to develop policies only as needed. The advantages are increased compliance because the quality system is considered by employees as a help rather than as a straightjacket. If a need for a policy is perceived, people will be willing to participate in the discussion and drafting process.

7.3.2. It is important to keep in mind that the policy is a contract (i.e., the quality agreed on by the function(s) signing the policy and the general requirements to achieve this quality). Policies should be kept general, as specific details or procedures should be developed in the form of standards and/or SOPs. The reason is that it will be possible to agree on a general goal or quality first. Then it will be far easier to discuss the procedures leading to this quality. It is also envisaged that standards and SOPs will be amended more often than policies, thus avoiding continual discussions on fundamental issues.

7.3.3. If required, policies can make provisions for the establishment of standards. This is always advisable when formal details or time frames must be determined or the quality can be specified in numbers. In this special case, a standard format has already been suggested and followed for all draft policies in this Quality Manual.

7.3.4. The quality system lives only if the policies are discussed and agreed on and standards and/or SOPs are developed to implement the generalized policies into wording for everyday work. What must happen to provide the feeling that this policy is appropriately met?

   While all policies agreed on should be completely implemented, there may be a need for additional standards or SOPs. This is acceptable, but make sure that they are integrated into the quality system and do not form an additional or contradictory system. Bear in mind that too many SOPs tend to lead to a decrease in compliance.

7.3.5. The highest hierarchical level affected should sign the policy in order to ensure not only willingness but also the power to take the necessary measures for implementation. Quality is a management task. It should not be left to an appointed quality manager without empowerment.

7.3.6. Audit procedures and feedback mechanisms are required if the quality system is to function appropriately. Otherwise it will be outdated the moment it is published. Although most organizations object to being controlled or inspected, audits are absolutely necessary to improve quality. Audits are usually accepted if it is made clear that the employees participate and that an agreed-on audit plan is used.

7.4. Policy Implementation

Give the name and title/function of authorized person(s) responsible for policy implementation. This must be in line with the authorization given on the cover sheet. Responsibility is not only for implementation of the respective policy but also for appropriate compliance with its provisions. Responsibility is more than signing the policy. The person responsible for implementation must see to it that adequate standards and/or SOPs are developed and that they are sufficiently detailed to ensure correct reflection in everyday work. Appropriate compliance means that the policy must be totally (not partially) implemented. The idea behind it should be followed rather than twisting its meaning.

7.5. Ruling Pertinent to Existing Procedures

If applicable, the impact of the document on already existing document(s) and vice versa should be discussed here. Such rulings include, for example, replacing, to be read in conjunction with, or implementation of another document issued at a higher decision level.

7.6. Ruling Applicable in the Case of Outsourcing

Generally speaking, all Regulatory Affairs work can be outsourced. Therefore, the quality system requires that adequate provisions are established for outsourcing regarding each work result or process governed by a policy. General guidelines are given in the policy on outsourcing (P-21).

7.7. Points to Consider During the Deployment of This Policy

For the application of this and all other policies, the general principles given above apply. They will not be repeated but will be referenced in the other policies. Additionally, special points to consider for the specific policy will be listed in section 7 of each policy, numbered according to the section of the policy they apply to (e.g., 7.3.1. will mean points to consider concerning section 3.1. of that particular policy).

A Policy of Department XYZ

02: Auditing and Compliance

Document Type:	Policy
Document Code:	02
	(enter company-specific code)
Title:	Auditing and Compliance
Date/Revision No.:	DD/MM/YY number xy
Scope:	Global
References:	(enter policies, standards, SOPs of your department/company, or other documents [e.g., guidelines] that should be considered in this context)
	1. Policy on Education/Training (P-12)
Authorization:	
	Signature of authorized person(s)
	Name of authorized person(s)
	Job title/Function of authorized person(s)
Issue Date:	DD/MM/YY
Implementation Date:	DD/MM/YY

1.   PURPOSE

This policy is a set of rules developed to specify principles for auditing and measures to ensure the compliance of Regulatory Affairs with the quality system.

2. DEFINITIONS

The key terms pertaining to this policy should be defined here. As there are no uniform and globally accepted definitions available, please develop your own definitions. In this way, the language of the staff of your organization can be incorporated. The term *employee* in this context means a person permanently employed by Regulatory Affairs, except if otherwise specified.

3. STATEMENT OF POLICY

This section covers the actual rulings that should be complied with when working according to the principles of the quality system. For the topic *auditing and compliance,* rulings should be available for the following items:

3.1.   All Regulatory Affairs employees are requested to comply with the provisions of the quality system as documented in the Quality Manual.

3.2.   All Regulatory Affairs employees are responsible for quality improvement and the development and maintenance of the quality system.

3.3.   All Regulatory Affairs employees are invited and allowed to participate in quality improvement activities.

3.4.   There will be a measurable, annual assessment of Regulatory Affairs's quality and, if required, a quality improvement process by setting defined quality objectives in the quality system.

3.5.   Education and training are vital to the quality improvement process and the development of and maintenance of the quality system (1). Quality training will be continuously available for all Regulatory Affairs employees.

3.6.   Emphasis must be on prevention rather than on control—quality built in, not inspected in.

3.7.   Regulatory Affairs will involve internal as well as external suppliers and customers in the development and maintenance of the quality system and the quality improvement process. Quality will be defined for each work result, for example, dossiers, and/or service or process by Policies. It will be defined in a measurable way by Standards and/or SOPs. For each work result, together with internal as well as external supplier(s) and customer(s), the necessary quality of input and output will be defined, focussing on the following:

- Specifications (e.g., critical items to be controlled)

- Limits or ranges of tolerance (e.g., in a dossier: number of missing pages)

- Frequency/extent of checks

- Documentation of check results

- Responsibility for such checks

- Necessity for duplication of checks by customer(s)

3.8.   A Quality Steering Committee will be responsible for decision making on quality management and the quality system and coordination of the quality improvement process.

3.9.   Continuous feedback is an important element of the system. Responsibility for feedback lies with each individual Regulatory Affairs employee.

3.10.   Once per year, an audit of all Regulatory Affairs Policies, Standards, and SOPs will be performed. It will be preannounced. An audit plan will be written and distributed in advance. The results of the audit will be documented in an audit report. Based on the principle of objectivity, there should be representation of supplier(s) and customer(s) during the audit. Additionally, a neutral third party should be present. If deficiencies are identified, adequate measures will be taken to improve the quality of the particular work result or process.

3.11.   The Quality Steering Committee will report to upper management on the status of the quality system once per year. The report will be signed by authorized person(s).

**4. RESPONSIBILITY FOR POLICY IMPLEMENTATION:
NAME AND TITLE/FUNCTION OF AUTHORIZED PERSON(S)**

5. RULING PERTINENT TO EXISTING PROCEDURES

6. RULING APPLICABLE IN THE CASE OF OUTSOURCING

**7. POINTS TO CONSIDER
DURING THE DEPLOYMENT OF THIS POLICY**

For the application of this and all other policies, general principles apply. (The general principles are given in 7.1–7.7 of P-01.)

7.2. Consider whether this policy should not also apply for nonpermanently employed persons who are replacing permanent employees for a lengthy period of time.

7.3.1. This sounds trivial, but the active involvement of everybody is the key to the success of any quality system. Use all of the resources and ideas available within Regulatory Affairs. Make sure that there is motivation and full commitment.

7.3.2./7.3.3. Ensure that there are multiple opportunities for suggestions of improvement. Do not rely on people to forward ideas proactively. Combining the annual evaluation of employee performance with the question for suggestions for improvement sounds like a good idea. Ask people to contribute ideas. The establishment of permanent quality circles may be another means of active involvement.

7.3.4. At least once per year, review the whole system. Think about the following:

- Is the quality as defined?

- Is this quality sufficient to meet the department's overall objectives?

- Are the customers satisfied?

Finally, reevaluate the system and set the objectives for next year's quality improvement measures as precisely as possible. Keep in mind that the goals should be realistic and measurable.

7.3.5. Continuous education is a key factor to ensure commitment. Policies, standards, and SOPs must become an integral part of everybody's daily life.

7.3.6. Control may produce anxiety among the employees, which may lead to even more mistakes and, consequently, higher rejection rates. Prevention only can produce sufficiently high quality with low rates of rejection. Control requires a lot of manpower and leads to lower commitment, whereas emphasis on prevention goes along with the empowerment of responsible people.

7.3.7. Quality depends, to a crucial extent, on internal and external suppliers. The definition of quality depends on what the customer perceives as quality. Both aspects should be borne in mind when defining quality for the department. This can be best achieved by integrating internal and external suppliers in the discussion process and, if possible, also reaching agreements. Quality is not a goal in itself, but the quality sufficient to reach departmental goals. Agreement on quality should be reached by a careful and detailed process of discussion. Beware of quick agreements or a high degree of harmony—they might be indicators of misunderstanding or failure to understand or accept the consequences. The process of defining measurable goals usually helps to avoid these pitfalls. Do not be discouraged if the discussion consumes a lot of time. It is time well spent.

Depending on the items, consider the extent of checks (e.g., by internal/external supplier, Regulatory Affairs department, internal/external customer). As a rule of thumb, checks for the quality of input should be left to the supplier. Make sure that the supplier operates according to agreement by a system of spot checks with frequency, the extent of tests adapted to the novelty of the agreement, and deviations identified.

7.3.9. Feedback concerning experience with the system and suggestions for improvement are of great importance. Responsibility lies with each employee.

7.3.10. Regular audits are required to check compliance with the elements of the quality system. This should be performed at least once per year. As people usually object to surprise audits, preannounce audits if at all possible. Good Regulatory Practice (GRP) is a new thing. Therefore, it is advisable to develop in advance an audit plan with the key elements to be checked. The development of such a plan is beneficial in that it helps to promote further understanding of the quality system. It is important to document the results in an audit report. Subsequent discussions should result, if required, in quality improvement measures. The Quality Steering Committee should carefully review the audit report in conjunction with the last audit report and the present quality goals to upper management.

An objective audit is hard to achieve. The suspicion of bias and prejudice prevails if audits are performed by only a single person. Thus, it is advisable to have audits performed by an audit team. Such teams should be composed of representatives of

- The audited unit
- The supplier(s)
- The customer(s)
- A member of a "neutral" institution

Make sure in advance, however, that the "neutral" member of the audit team is recognized as being neutral. Otherwise, suspicion will produce resistance.

A Policy of Department XYZ

03: Application for Clinical Trial License

Document Type: Policy

Document Code: 03

                        (enter company-specific code)

Title: Application for Clinical Trial License

Date/Revision No.:   DD/MM/YY number xy

Scope: Global

References: (enter policies, standards, SOPs of your department/company,
                        or other documents [e.g., guidelines] that should be consid-
                        ered in this context)

                        1. Policy on Submission (P-27)

                        2. Standard for U.S. Application for Clinical Trial
                           License: IND Content and Format (S-03.01)

Authorization: _____

                        Signature of authorized person(s)

                        _____

                        Name of authorized person(s)

                        _____

                        Job title/Function of authorized person(s)

Issue Date: DD/MM/YY

Implementation Date: DD/MM/YY

---

1. PURPOSE

This policy is a set of rules developed to govern the generation of an Application for a Clinical Trial License by Regulatory Affairs worldwide.

2. DEFINITIONS

The key terms pertaining to this policy should be defined here. As there are no uniform and globally accepted definitions available, please develop your own definitions. In this way, the language of the staff of your organization can be incorporated. The term *Application for a Clinical Trial License* signifies a notification or application to a Regulatory Body with the purpose of starting local trials in humans.

3. STATEMENT OF POLICY

This section covers the actual rulings that should be complied with when working according to the principles of the quality system. For the topic *Application for a Clinical Trial License,* rulings should be available for the following items:

3.1. For each trial in humans with the company's developmental and/or already marketed medicinal product(s), the applicable regulatory requirements will be adequately met.

3.2. Responsibility and a single point of reference for the generation and documentation of the Application for a Clinical Trial License is with the assigned Regulatory Affairs manager. Assignments are documented in the submission assignments listing (see the policy on submission [1]).

3.3. The Application for a Clinical Trial License is generated after the official decision to request a Clinical Trial License in a specific country.

3.4. The Application for a Clinical Trial License will be generated in a timely fashion, as per the policy on submission (1).

3.5. Each organizational unit responsible for generating the Application for a Clinical Trial License provides and maintains a standard for format and structure (2).

4. RESPONSIBILITY FOR POLICY IMPLEMENTATION: NAME AND TITLE/FUNCTION OF AUTHORIZED PERSON(S)

5. RULING PERTINENT TO EXISTING PROCEDURES

6. RULING APPLICABLE IN THE CASE OF OUTSOURCING

7. POINTS TO CONSIDER DURING THE DEPLOYMENT OF THIS POLICY

For the application of this and all other policies, general principles apply. (The general principles are given in 7.1–7.7 of P-01.)

7.2.   Additional, related documentation may include the Clinical Trial Exemption (CTX), the Clinical Trial Certificate (CTC), and the Investigational New Drug (IND) application.

7.3.1. For applicable regulatory requirements, see national regulations and guidelines. Adequately meeting these requirements will mean fulfilling the requirements from a scientific and/or strategic point of view. Above all, the company must ascertain that the medicinal product is safe for human use in the intended dose range and under the intended conditions of the planned clinical trial.

7.3.4. Define the time frame for your organization. A suggested time frame is within one month from the official clinical trial submission decision date.

7.3.5. Develop a standard for the format and the structure of an application for a clinical trial authorization in countries relevant for your company. For INDs in the United States, see S-03.01.

A Policy of Department XYZ

04: Application for Marketing Authorization

Document Type:	Policy
Document Code:	04
	(enter company-specific code)
Title:	Application for Marketing Authorization
Date/Revision No.:	DD/MM/YY number xy
Scope:	Global
References:	(enter policies, standards, SOPs of your department/company, or other documents [e.g., guidelines] that should be considered in this context)

1. Policy on Submission (P-27)

2. Standard for EU Application for Marketing Authorization: Chemical Active Substance(s) (S-04.01)

3. Standard for EU Application for Marketing Authorization: Biological(s), Part II (S-04.02)

4. Standard for U.S. Application for Marketing Authorization (S-04.03)

Authorization:	
	Signature of authorized person(s)
	Name of authorized person(s)
	Job title/Function of authorized person(s)
Issue Date:	DD/MM/YY
Implementation Date:	DD/MM/YY

1. PURPOSE

This policy is a set of rules developed to govern the generation of an Application for Marketing Authorization by Regulatory Affairs worldwide.

2. DEFINITIONS

The key terms pertaining to this policy should be defined here. As there are no uniform and globally accepted definitions available, please develop your own definitions. In this way, the language of the staff of your organization can be incorporated. The term *Application for Marketing Authorization* signifies a notification or application to a Regulatory Body with the purpose of placing a medicinal product on the market.

3. STATEMENT OF POLICY

This section covers the actual rulings that should be complied with when working according to the principles of the quality system. For the topic *Application for Marketing Authorization,* rulings should be available for the following items:

3.1.   For each of the company's developmental and/or already marketed medicinal product(s), the applicable regulatory requirements will be adequately met.

3.2.   Responsibility and a single point of reference for the generation and documentation of the Application for Marketing Authorization is with the assigned Regulatory Affairs manager. Assignments are documented in the submission Assignments Listing (see the policy on submission [1]).

3.3.   The Application for Marketing Authorization is generated after the official decision to request a marketing authorization in a specific country.

3.4.   The Application for Marketing Authorization will be generated in a timely fashion, as per the policy on submission (1).

3.5.   Each organizational unit responsible for generating the Application for Marketing Authorization provides and maintains a standard for format and structure (2,3,4).

4. RESPONSIBILITY FOR POLICY IMPLEMENTATION: NAME AND TITLE/FUNCTION OF AUTHORIZED PERSON(S)

5. RULING PERTINENT TO EXISTING PROCEDURES

6. RULING APPLICABLE IN THE CASE OF OUTSOURCING

7. POINTS TO CONSIDER DURING THE DEPLOYMENT OF THIS POLICY

For the application of this and all other policies, general principles apply. (The general principles are given in 7.1–7.7 of P-01.)

7.3.1. For applicable regulatory requirements, see national regulations and guidelines. Adequately meeting these requirements will mean fulfilling the requirements from a scientific and/or strategic point of view. Above all, the company must be sure that the medicinal product is of adequate quality, efficacy, and safety.

7.3.4. Define the time frame for your organization. Suggested time frame is within one to two months from official decision to request a Marketing Authorization in a specific country.

7.3.5. Develop a standard for format and structure. For the EU and U.S. Application for Marketing Authorization, see Standards S-04.01–03 (2,3,4).

A Policy of Department XYZ

05: Archiving Management

Document Type:	Policy
Document Code:	05
	(enter company-specific code)
Title:	Archiving Management
Date/Revision No.:	DD/MM/YY number xy
Scope:	Global
References:	(enter policies, standards, SOPs of your department/company, or other documents [e.g., guidelines] that should be considered in this context)
	1. Standard for Regulatory Document Types (S-10.01)
Authorization:	
	Signature of authorized person(s)
	Name of authorized person(s)
	Job title/Function of authorized person(s)
Issue Date:	DD/MM/YY
Implementation Date:	DD/MM/YY

1. PURPOSE

This policy is a set of rules developed to govern the identification, retention, storage, protection, and disposal of the records of Regulatory Affairs world-wide. The objective of this policy is to

- Eliminate unnecessary records

- Ensure that records worth archiving are defined

- Specify archiving rules for records

- Ensure compliance with legal and business obligations by adequate archiving of records

- Avoid redundant or excessively long archiving of records in order to save costs (e.g., rent, personnel, and handling)

2. DEFINITIONS

The key terms pertaining to this policy should be defined here. As there are no uniform and globally accepted definitions available, please develop your own definitions. In this way, the language of the staff of your organization can be incorporated.

- The term *record* includes written documents and records maintained on microfilm, optical disc, magnetic tape, or other media. Records are kept in official archives (or centralized repositories) and records are kept in individual files.

- *Archiving management* refers to identification, retention, storage, protection, and disposal of the records of Regulatory Affairs, including records on the developmental research of medicinal products and already marketed medicinal products.

3. STATEMENT OF POLICY

This section covers the actual rulings that should be complied with when working according to the principles of the quality system. For the topic *archiving management,* rulings should be available for the following items:

3.1.   Records that need archiving should be created and distributed strictly as needed.

3.2.   A standard (1) will be developed to define and identify record types for which Regulatory Affairs is responsible for the archiving of the original or the master copy on behalf of the company because of legal and/or business obligations and other record types that are created or received during everyday work in Regulatory Affairs. Special attention should also be given to records created by electronic storage media (e.g., word processing systems, E-mail, or databases). The standard for these record types will specify retention times and requirements (e.g., storage, need for safety copy and access limitations) and, if necessary, identify locations and contact persons.

3.3.   In order to facilitate access to relevant information on registration status and submissions worldwide, Regulatory Affairs maintains an archive of all Regulatory Affairs documents.

4.   RESPONSIBILITY FOR POLICY IMPLEMENTATION: NAME AND TITLE/FUNCTION OF AUTHORIZED PERSON(S)

5.   RULING PERTINENT TO EXISTING PROCEDURES

6.   RULING APPLICABLE IN THE CASE OF OUTSOURCING

7.   POINTS TO CONSIDER DURING THE DEPLOYMENT OF THIS POLICY

For the application of this and all other policies, general principles apply. (The general principles are given in 7.1–7.7 of P-01.)

7.2.1.   As the decision to create and distribute records lies within the responsibility of the individual author of a record, it is important that this policy is clearly understood by all Regulatory Affairs employees. However, they should be encouraged to consider whether they actually need to receive copies regularly or need to be included in standard distribution lists.

7.3.1.   It is important to agree on the mentioned types of record lists. You will find that agreeing on the terms and definitions and, if required, examples, is a lengthy but rewarding process. It is especially beneficial for large organizations. This discussion process will help you to use one terminology and greatly reduce misunderstandings. Review the applicable legal and/or business obligations and then list the retention times and requirements. It is suggested that the legal department be involved at this stage to ensure that the archive location(s), personnel, and safety copies meet the requirements. In case of more than one archive location, it is necessary to identify the location(s) and the contact person(s) to facilitate user access to records or copies of records.

7.3.2. Develop a standard for Regulatory Document Types, based on the suggested S-10.01. It is advisable that Regulatory Affairs maintain a (centralized) archive for easy access to relevant information on key Regulatory Affairs activities (e.g., registration status, information on submissions), as this information is typically needed within hours (e.g., by upper management or by Drug Safety).

A Policy of Department XYZ

06: Change Alert/Authorization Process

Document Type:	Policy
Document Code:	06
	(enter company-specific code)
Title:	Change Alert/Authorization Process
Date/Revision No.:	DD/MM/YY number xy
Scope:	Global
References:	(enter policies, standards, SOPs of your department/company, or other documents [e.g., guidelines] that should be considered in this context)
	1. Departmental operational procedures
	2. Policy on Global Dossier (P-15)
	3. Policy on Labeling (P-20)
	4. Policy on Information Management (P-17)
	5. Policy on Dossier (P-11)
	6. Policy on Submission (P-27)
	7. Policy on Contact with Regulatory Body (P-08)
	8. Policy on Contact Report (P-07)
	9. Policy on Archiving Management (P-05)

Authorization:

_____
Signature of authorized person(s)

_____
Name of authorized person(s)

_____
Job title/Function of authorized person(s)

Issue Date: DD/MM/YY
_____

Implementation Date: DD/MM/YY
_____

1. PURPOSE

This policy is a set of rules developed to specify the principles applying to changes, namely authorization process and change alert procedures, for Regulatory Affairs worldwide.

2. DEFINITIONS

The key terms pertaining to this policy should be defined here. As there are no uniform and globally accepted definitions available, please develop your own definitions. In this way, the language of the staff of your organization can be incorporated.

- The term *change* in this policy signifies any changes or variations concerning the company's substances or medicinal products worldwide. This includes, for example, changes in starting materials, manufacturing process/equipment/batch size/site, dosage form, packaging, controls, patient leaflet, labeling, or outer packaging. If subject to submission to the Regulatory Bodies, any additional information, such as advertising/promotional material, is also included.

- The terms *substance* and *medicinal product* in this policy apply both during the development life cycle and the market life cycle. They also apply whenever the company has legal responsibility, be it as the contract manufacturer, the person responsible for bringing the product into the market, the partner in co-marketing, the licensor or licensee, the holder of a Drug Master File (DMF) or of a letter of authorization, or as a partner in joint development.

3. STATEMENT OF POLICY

This section covers the actual rulings that should be complied with when working according to the principles of the quality system. For the topic *change alert/authorization process,* rulings should be available for the following items:

3.1. No change concerning the company's substances or medicinal products may be implemented after the submission of the dossier in at least one country without approval by Regulatory Affairs. For change alert, a special procedure must be in place (1).

3.2. Responsibility for input concerning regulatory requirements is with Regulatory Affairs.

3.3.   Legal and business obligations will be observed. Special attention will be given to the registration status of the product and the regulatory requirements for variations in the countries concerned.

3.4.   If required, the global dossier (2) and labeling (3) will be adapted by Regulatory Affairs in a timely fashion.

3.5.   Regulatory Bodies will be informed of changes appropriately and in a timely fashion. Policies on information management (4), dossier (5), submission (6), contact with regulatory bodies (7), contact reports (8), and archiving management (9) apply.

4.   RESPONSIBILITY FOR POLICY IMPLEMENTATION: NAME AND TITLE/FUNCTION OF AUTHORIZED PERSON(S)

5.   RULING PERTINENT TO EXISTING PROCEDURES

6.   RULING APPLICABLE IN THE CASE OF OUTSOURCING

7.   POINTS TO CONSIDER DURING THE DEPLOYMENT OF THIS POLICY

For the application of this and all other policies, general principles apply. (The general principles are given in 7.1–7.7 of P-01.)

7.2.   This policy deals exclusively with the regulatory aspect of changes or variations. Total change management lies beyond the scope of this policy. Note that the definition of life cycle may vary from country to country. With regard to liability claims, however, the life cycle, as a rule of thumb, is assumed to end 30 years after withdrawal from the market.

7.3.1.   As there are different regulations in countries for variations (e.g., requiring notice, approval, or marketing authorization), every internal application to change a product must be carefully evaluated by Regulatory Affairs on a case-by-case basis. If no in-house change authorization procedure is in place, it should be established. The following functions should be involved: Manufacturing, Quality Control, Drug Safety, Marketing, Regulatory Affairs, and others as needed. Regulatory Affairs should also be involved in determining the distribution per case. For a change alert, a special procedure that ensures adequate quick action should be in place.

7.3.3. Consider whether documents for regulatory purposes must be amended. Have previous versions been submitted to Regulatory Bodies? Which regulatory requirements apply for notification/ approval of variations in the countries where the change is projected? What is the impact on other countries (e.g., free sales certificate or reimport)?

7.3.4. If the change is approved, make sure that the required documents for regulatory purposes are written and included in the global dossier in a timely fashion. This applies also to labeling changes.

7.3.5. For changes that require approval or a new marketing authorization, no change should be carried out until the official permission has been received.

A Policy of Department XYZ

07: Contact Report

Document Type:	Policy
Document Code:	07
	(enter company-specific code)
Title:	Contact Report
Date/Revision No.:	DD/MM/YY number xy
Scope:	Global
References:	(enter policies, standards, SOPs of your department/company, or other documents [e.g., guidelines] that should be considered in this context)
	1. Policy on Contact with Regulatory Body (P-08)
	2. Standard for Regulatory Body Contact Report (S-07.01)
	3. Policy on Inspection (P-19)
	4. Policy on Archiving Management (P-05)
Authorization:	
	Signature of authorized person(s)
	Name of authorized person(s)
	Job title/Function of authorized person(s)
Issue Date:	DD/MM/YY
Implementation Date:	DD/MM/YY

1. PURPOSE

This policy is a set of rules developed to specify the principles applying to the contact with Regulatory Bodies worldwide.

2. DEFINITIONS

The key terms pertaining to this policy should be defined here. As there are no uniform and globally accepted definitions available, please develop your own definitions. In this way, the language of the staff of your organization can be incorporated.

- The term *contact report* signifies a report on contact with Regulatory Bodies in a standardized format.

- The term *contact with Regulatory Body* signifies any communication between Regulatory Affairs and a Regulatory Body, directly, by phone, E-mail, fax, telex, letter, or during a personal meeting.

3. STATEMENT OF POLICY

This section covers the actual rulings that should be complied with when working according to the principles of the quality system. For the topic *contact report,* rulings should be available for the following items:

3.1.   Responsibility for contact with Regulatory Bodies lies with Regulatory Affairs (1).

3.2.   Within 24 hours following each contact with a Regulatory Body, a contact report must be written by Regulatory Affairs (unless a self-explanatory document was written/received).

3.3.   The format and content of the contact report are defined (2). For inspections, see also the policy on inspection (3).

3.4.   Distribution of the contact report will be to appropriate persons, and the policy on archiving applies (4).

**4. RESPONSIBILITY FOR POLICY IMPLEMENTATION:
    NAME AND TITLE/FUNCTION OF AUTHORIZED PERSON(S)**

5. RULING PERTINENT TO EXISTING PROCEDURES

6. RULING APPLICABLE IN THE CASE OF OUTSOURCING

**7. POINTS TO CONSIDER
    DURING THE DEPLOYMENT OF THIS POLICY**

For the application of this and all other policies, general principles apply.
(The general principles are given in 7.1–7.7 of P-01.)

7.3.1. The responsibility for contact with Regulatory Bodies is a topic under debate in some companies. In U.S.–based companies, contact is
as suggested by this policy, through Regulatory Affairs. In other companies company experts contact the Regulatory Body exclusively or
in addition to Regulatory Affairs. However, it is strongly recommended for the sake of clear and good communication to identify a
single contact person. Ideally, a specific person should be named for
each project. This person should be the most knowledgeable about
the dossier and the location of information in it and have profound
knowledge of the ways and procedures of the Regulatory Body in
question. This will usually be the Regulatory Affairs manager. However, in case of a lawsuit, for example, delegate responsibility for
contact with Regulatory Bodies to a lawyer because of the specialized legal wording and procedures involved.

7.3.2. It is important to write down the summary results of the contact as
soon as possible in order not to omit important details or to include
too much interpretation.

7.3.3. The format and minimum content should be standardized. Thus, a
standard is suggested (see S-07.01).

7.3.4. It is suggested to define distribution lists, if required, on a project-by-
project basis. Contact reports must be archived, as they document an
official contact with a Regulatory Body.

A Policy of Department XYZ

08: Contact with Regulatory Body

Document Type:	Policy
Document Code:	08
	(enter company-specific code)
Title:	Contact with Regulatory Body
Date/Revision No.:	DD/MM/YY number xy
Scope:	Global
References:	(enter policies, standards, SOPs of your department/company, or other documents [e.g., guidelines] that should be considered in this context)
	1. Policy on Information Management (P-17)
	2. Policy on Crisis Management (P-09)
	3. Policy on Change Alert/Authorization Process (P-06)
	4. Policy on Contact Report (P-07)
	5. Policy on Dossier (P-11)
	6. Policy on Submission (P-27)
	7. Policy on Periodic Safety Update Report (P-22)

Authorization:

_____

Signature of authorized person(s)

_____

Name of authorized person(s)

_____

Job title/Function of authorized person(s)

Issue Date: DD/MM/YY
_____

Implementation Date: DD/MM/YY
_____

1. PURPOSE

This policy is a set of rules developed to specify the principles applying to the contact with Regulatory Bodies by Regulatory Affairs worldwide.

2. DEFINITIONS

The key terms pertaining to this policy should be defined here. As there are no uniform and globally accepted definitions available, please develop your own definitions. In this way, the language of the staff of your organization can be incorporated. The term *contact with Regulatory Body* signifies any communication between Regulatory Affairs and a Regulatory Body, directly, by phone, E-mail, fax, telex, letter, or during a personal meeting.

3. STATEMENT OF POLICY

This section covers the actual rulings that should be complied with when working according to the principles of the quality system. For the topic *contact with Regulatory Body*, rulings should be available for the following items:

3.1. Responsibility for contact with Regulatory Bodies lies with Regulatory Affairs. No contact by the company personnel, experts, or consultants acting on the company's behalf may be made without prior knowledge and agreement by Regulatory Affairs. However, Regulatory Affairs may appoint qualified person(s) to act on its behalf.

3.2. The Regulatory Affairs representative may, if necessary, ask to be accompanied by other persons (e.g., experts from the scientific disciplines or legal). Responsibility for the selection of appropriate accompanying persons and the preparation of the meeting is with Regulatory Affairs.

3.3. Information communicated to/from the Regulatory Bodies will be handled adequately. Policies on information management (1), crisis management (2), change alert/authorization process (3), contact report (4), dossier (5), submission (6), and the periodic Safety Update Report (7) also apply.

3.4. A contact report will be written by Regulatory Affairs for each contact with a Regulatory Body (4) (unless a self-explanatory document has been written/received).

**4. RESPONSIBILITY FOR POLICY IMPLEMENTATION:
NAME AND TITLE/FUNCTION OF AUTHORIZED PERSON(S)**

5. RULING PERTINENT TO EXISTING PROCEDURES

6. RULING APPLICABLE IN THE CASE OF OUTSOURCING

**7. POINTS TO CONSIDER
DURING THE DEPLOYMENT OF THIS POLICY**

For the application of this and all other policies, general principles apply. (The general principles are given in 7.1–7.7 of P-01.)

7.3.1. The responsibility for contact with Regulatory Bodies is a topic under debate in some companies. In U.S.–based companies, contact is as suggested by this policy, through Regulatory Affairs. In other companies, company experts contact the Regulatory Body exclusively or in addition to Regulatory Affairs. However, it is strongly recommended for the sake of clear and good communication to identify a single contact person. Ideally, a specific person should be named for each project. This person should be the most knowledgeable about the dossier and the location of information in it and have profound knowledge of the ways and procedures of the Regulatory Body in question. This will usually be the Regulatory Affairs manager. However, in case of a lawsuit, it is recommended to delegate responsibility for contact with Regulatory Bodies to a lawyer because of the specialized legal wording and procedures involved.

7.3.2. If required, inside and/or outside experts can also be consulted. Regulatory Affairs should play a key role in deciding when a hearing can be beneficial.

7.3.4. Documenting the contact by writing a contact report for future reference is important. Therefore, an individual policy on contact reports is suggested.

A Policy of Department XYZ

09: Crisis Management

Document Type:	Policy
Document Code:	09
	(enter company-specific code)
Title:	Crisis Management
Date/Revision No.:	DD/MM/YY number xy
Scope:	Global
References:	(enter policies, standards, SOPs of your department/company, or other documents [e.g., guidelines] that should be considered in this context)
	1. Policy on Information Management (P-17)
Authorization:	
	Signature of authorized person(s)
	Name of authorized person(s)
	Job title/Function of authorized person(s)
Issue Date:	DD/MM/YY
Implementation Date:	DD/MM/YY

1.   PURPOSE

This policy is a set of rules developed to specify the principles governing crisis management in Regulatory Affairs worldwide.

2.   DEFINITIONS

The key terms pertaining to this policy should be defined here. As there are no uniform and globally accepted definitions available, please develop your own definitions. In this way, the language of the staff of your organization can be incorporated. The term *crisis* in the context of this policy signifies any situation with a high potential of danger or damage to the company's reputation, substances, or medicinal products, or personnel within the scope of responsibilities of Regulatory Affairs or closely connected with its function or directly affecting Regulatory Affairs personnel in their function. Usually, a crisis will appear unforeseen or suddenly, or a situation that appeared to be under control will worsen within a short period of time. Typically, a crisis involves both time pressure and emotional pressure. A quick and well-planned action will be required to improve the situation and/or prevent further damage.

3.   STATEMENT OF POLICY

This section covers the actual rulings that should be complied with when working according to the principles of the quality system. For the topic *crisis management,* rulings should be available for the following items:

3.1.   For situations where Regulatory Affairs is informed of possible danger to the company's reputation, substances, or medicinal products but is not immediately responsible or concerned, see the policy on information management (1).

3.2.   In a crisis, the responsibilities of Regulatory Affairs functions remain as defined. However, in a crisis, every employee should act to the best of his or her abilities to help solve the crisis or prevent or limit damage as far as possible.

3.3.   Possible situations of crisis will be listed and then explored on a generic basis in order to reach a good understanding of the risks involved. Crisis scenarios will be developed. Appropriate measures will be put in place to prevent the advent of such a crisis situation. Action plans will be developed as needed in order to be available in case of crisis.

**4.   RESPONSIBILITY FOR POLICY IMPLEMENTATION:
NAME AND TITLE/FUNCTION OF AUTHORIZED PERSON(S)**

5.   RULING PERTINENT TO EXISTING PROCEDURES

6.   RULING APPLICABLE IN THE CASE OF OUTSOURCING

**7.   POINTS TO CONSIDER
DURING THE DEPLOYMENT OF THIS POLICY**

For the application of this and all other policies, general principles apply. (The general principles are given in 7.1–7.7 of P-01.)

7.3.3   Prepare a list of potential crises for your organization. For example,

- Severe product-related problems with regard to safety, quality, or efficacy

- Intervention by a Regulatory Body (e.g., revocation of marketing authorization, product recall, cancelling of indications, addition of contraindications, interactions, warning statements) because of safety, quality, or efficacy concerns

- Failure in performance by Regulatory Affairs (e.g., missing the deadline for the application for renewal of a marketing authorization with the risk of losing marketing authorization)

- Inability to carry out a regulatory function appropriately (e.g., absence of key personnel, breakdown of electronic data processing [EDP] systems or other machines, power failure, or sudden need for relocation to nonfully equipped site)

- Loss/theft of confidential information (e.g., theft of single documents or dossiers within Regulatory Affairs or in the mail), or a breach of a confidentiality agreement by Regulatory Affairs employees

- Loss of records or documents on Regulatory Affairs premises by accident (e.g., fire, water damage, breakdown of EDP systems)

- Imminent danger to health and/or life of Regulatory Affairs personnel (e.g., fire or heart attack)

Establish alarm and/or action plans for each of these situations and include these in regular training measures.

A Policy of Department XYZ

10: Documents for Regulatory Purposes

Document Type:	Policy
Document Code:	10
	(enter company-specific code)
Title:	Documents for Regulatory Purposes
Date/Revision No.:	DD/MM/YY number xy
Scope:	Global
References:	(enter policies, standards, SOPs of your department/company, or other documents [e.g., guidelines] that should be considered in this context)
	1. Policy on Global Dossier (P-15)
	2. Policy on Regulations and Guidelines (P-25)
Authorization:	
	Signature of authorized person(s)
	Name of authorized person(s)
	Job title/Function of authorized person(s)
Issue Date:	DD/MM/YY
Implementation Date:	DD/MM/YY

1.   PURPOSE

This policy is a set of rules developed to specify the principles governing the standardization of documents for regulatory purposes and the general input of Regulatory Affairs concerning the development and maintenance of dossiers, submissions, and, if applicable, global dossiers, for the company's medicinal products.

2.   DEFINITIONS

The key terms pertaining to this policy should be defined here. As there are no uniform and globally accepted definitions available, please develop your own definitions. In this way, the language of the staff of your organization can be incorporated.

- The term *document for regulatory purposes* signifies any document that is intended for regulatory purposes (e.g., application for clinical trial authorization or application for marketing authorization).

- The term *internal company standard for documents for regulatory purposes* signifies a generic (not product-specific) document that gives the company's evaluation and summary of actual regulatory requirements for the format and content of documents for regulatory purposes and cross-references regulations and guidelines.

3.   STATEMENT OF POLICY

This section covers the actual rulings that should be complied with when working according to the principles of the quality system. For the topic *documents for regulatory purposes,* rulings should be available for the following items:

3.1.   The responsibility for input concerning documents for regulatory purposes with regard to regulatory know-how is with Regulatory Affairs.

3.2.   Regulatory Affairs input will be via internal company standards or via the project team for product-specific requirements.

3.3.   Internal company standards reflect the current state of the company's evaluation of actual regulations and guidelines.

3.4.   Decision making on internal standards of the company for documents for regulatory purposes is a joint process between Regulatory

Affairs and other concerned department(s) and/or scientific discipline(s). The coordination and maintenance of internal company standards for documents for regulatory purposes is the responsibility of Regulatory Affairs.

3.5.   Internal company standards for documents for regulatory purposes are updated continuously as needed and made available by Regulatory Affairs to other concerned departments and/or scientific disciplines.

3.6.   Product-specific documents for regulatory purposes meeting the internal standards of the company form the basis for dossiers.

4.   RESPONSIBILITY FOR POLICY IMPLEMENTATION: NAME AND TITLE/FUNCTION OF AUTHORIZED PERSON(S)

5.   RULING PERTINENT TO EXISTING PROCEDURES

6.   RULING APPLICABLE IN THE CASE OF OUTSOURCING

7.   POINTS TO CONSIDER DURING THE DEPLOYMENT OF THIS POLICY

For the application of this and all other policies, general principles apply. (The general principles are given in 7.1–7.7 of P-01.)

7.3.2. The requirements for documentation should consider general aspects (i.e., applicability to all or to a range of medicinal products) and product-specific requirements.

7.3.3. Much of the time and repetition of efforts in projects can be avoided by considering the requirements in advance and documenting the company's evaluation or agreement in written form. Of course, Regulatory Affairs must closely monitor evolving guidelines that may impact internal company standards. Even though many guidance documents are not legally binding, such as recommendations, notes for guidance, and points to consider (also in draft stage!), these documents form the actual state of the art and will, therefore, be used as a basis for decisions by Regulatory Bodies.

7.3.4. The coordination of internal company standards is typically the responsibility of Regulatory Affairs because Regulatory Affairs is the internal customer of documents for regulatory purposes.

7.3.6. It must be ensured that all documents for regulatory purposes meet internal company standards. The various scientific disciplines must guarantee that their documents meet such standards. Otherwise, a time-consuming, detailed check by Regulatory Affairs will be necessary to guarantee the quality of dossier(s).

A Policy of Department XYZ

11: Dossier

Document Type:	Policy
Document Code:	11
	(enter company-specific code)
Title:	Dossier
Date/Revision No.:	DD/MM/YY number xy
Scope:	Global
References:	(enter policies, standards, SOPs of your department/company, or other documents [e.g., guidelines] that should be considered in this context)
	1. Policy on Global Dossier (P-15)
	2. Policy on Documents for Regulatory Purposes (P-10)
	3. Policy on Project Assignments (P-23)
	4. Policy on Submission (P-27)
	5. Departmental operational procedures
Authorization:	
	Signature of authorized person(s)
	Name of authorized person(s)
	Job title/Function of authorized person(s)
Issue Date:	DD/MM/YY
Implementation Date:	DD/MM/YY

1. PURPOSE

This policy is a set of rules developed to govern the generation of a dossier by Regulatory Affairs worldwide.

2. DEFINITIONS

The key terms pertaining to this policy should be defined here. As there are no uniform and globally accepted definitions available, please develop your own definitions. In this way, the language of the staff of your organization can be incorporated.

- The term *dossier* signifies a compilation of documents for a specific regulatory purpose (e.g., application for clinical trial authorization or application for marketing authorization) in a specified country(ies) for a developmental or already marketed medicinal product in a structured form (i.e., submission-like). If applicable, it is a subset of the global dossier. The dossier is the basis for the submission(s).

- The term *global dossier* signifies a compilation of all documents required for international regulatory purpose(s) for a developmental or already marketed medicinal product. It is maintained continuously throughout the life cycle of the medicinal product and serves as a repository for the generation of dossiers and submissions.

- The term *submission* signifies a country-specific compilation of documents for a specific regulatory purpose (e.g., application for clinical trial authorization or application for marketing authorization) for a developmental or already marketed medicinal product in a structured form according to national regulatory requirements. It is based on the dossier, or, if applicable, the global dossier. It may contain additional national documents (e.g., national leaflets or application forms).

3. STATEMENT OF POLICY

This section covers the actual rulings that should be complied with when working according to the principles of the quality system. For the topic *dossier*, rulings should be available for the following items:

3.1.   If applicable, the dossier is a subset of the global dossier (1). The structure, format, and content of elements are defined by internal company standards (2).

3.2.   Responsibility and a single point of reference for the generation and documentation of the dossier is with the assigned Regulatory Affairs manager. Assignments are documented in the listing of project assignments (3).

3.3.   The dossier is generated after the official decision to submit an application for a medicinal product in at least one country.

3.4.   The dossier will be made available in a timely fashion to the Regulatory Affairs unit responsible for the submission (4).

3.5.   Each organizational unit responsible for generating a dossier provides a written procedure for processes and responsibilities (5).

4.   RESPONSIBILITY FOR POLICY IMPLEMENTATION: NAME AND TITLE/FUNCTION OF AUTHORIZED PERSON(S)

5.   RULING PERTINENT TO EXISTING PROCEDURES

6.   RULING APPLICABLE IN THE CASE OF OUTSOURCING

7.   POINTS TO CONSIDER DURING THE DEPLOYMENT OF THIS POLICY

For the application of this and all other policies, general principles apply. (The general principles are given in 7.1–7.7 of P-01.)

7.1.   One of the most important Regulatory Affairs functions is to generate dossiers for submission for regulatory purposes (e.g., to obtain a clinical trial license, a marketing authorization, or a renewal of a marketing authorization). The technical part of this process is reflected by policies on global dossier, dossier, and submission.

7.2.   For companies continuously maintaining a global dossier, this may be identical with the dossier. However, most companies provide local Regulatory Affairs units not with the complete documentation but with only the documentation subset relevant for the intended submission. Usually, the dossier is provided in a submission-like structure to facilitate generation of the submission. Based on your procedures, define whether the dossier will or will not be paginated and referenced. The submission typically contains additional locally required documents (e.g., application forms and national leaflet texts).

7.3.1.   In order to guarantee the quality of the dossier, the various scientific disciplines should confirm that their documents for regulatory purposes are generated according to internal company standards. Otherwise a time- and capacity-consuming check by Regulatory Affairs becomes necessary for all documents.

7.3.4.   Timelines apply only if separate Regulatory Affairs units are responsible for the generation of the dossier and the submission. If this is not the case, reflect this in your in-house procedure. Define the time frames for your organization. A suggested time frame is within one month from the receipt of the last document at Regulatory Affairs.

7.3.5.   Develop SOP(s) for generating a dossier.

A Policy of Department XYZ

12: Education/Training

Document Type:	Policy
Document Code:	12
	(enter company-specific code)
Title:	Education/Training
Date/Revision No.:	DD/MM/YY number xy
Scope:	Global
References:	(enter policies, standards, SOPs of your department/company, or other documents [e.g., guidelines] that should be considered in this context)
	1. Policy on Information Management (P-17)
Authorization:	
	Signature of authorized person(s)
	Name of authorized person(s)
	Job title/Function of authorized person(s)
Issue Date:	DD/MM/YY
Implementation Date:	DD/MM/YY

1.   PURPOSE

This policy is a set of rules developed to specify the principles that apply to the training and education of Regulatory Affairs staff worldwide. As people are the most important resource in ensuring high quality work in Regulatory Affairs, it must be recognized that Regulatory Affairs has an obligation to manage the training and education of its staff together with the personnel department. Regulatory Affairs staff must develop and maintain expertise in a wide range of regulatory-related activities and have a thorough understanding of current regulations governing worldwide medicinal product development. As only a few traditional educational facilities offer Regulatory Affairs training, it is important to organize in-house training. Staff should be sent to extramural training courses for general as well as specialized education (e.g., trainings, certification programs organized by Regulatory Affairs professional societies).

2.   DEFINITIONS

The key terms pertaining to this policy should be defined here. As there are no uniform and globally accepted definitions available, please develop your own definitions. In this way, the language of the staff of your organization can be incorporated. The terms *training* and *education* in this policy mean both internally and externally organized, theoretical, and/or practical measures to ensure achievement and/or maintenance of a high standard of knowledge, skills, and experience required within Regulatory Affairs.

3.   STATEMENT OF POLICY

This section covers the actual rulings that should be complied with when working according to the principles of the quality system. For the topic *education/training,* rulings should be available for the following items:

3.1.   For newcomers, attendance at a seminar that provides comprehensive information about the company—its organization, areas of business, and medicinal products—is obligatory with the aim to obtain knowledge about the company and the organization.

3.2.   A mentor from the Regulatory Affairs department is recommended for newcomers for their first year.

3.3.   For newcomers, feedback should be solicited from/to the department head after 3, 6, and 12 months.

3.4.   A general education plan should be established for newcomers (to be modified according to his or her education and training and the function).

3.5.   Attendance is required at presentations of departments/scientific disciplines with which Regulatory Affairs frequently interacts (e.g., Drug Safety, Clinical Pharmacology, Biometry, Project Management, Marketing) with the aim to obtain insight into structure and working methods and to establish contacts that should aid future collaboration.

3.6.   Regulatory Affairs will hold various trainings and further educational seminars at regular intervals as a source of knowledge of and training in Regulatory Affairs and to enable the acquisition of the expertise needed for the work.

3.7.   For extramural seminars and congresses, see the policy on information management (1).

3.8.   A trip report should be written for each extramural event and distributed to the interested parties.

3.9   An evaluation as to the quality and the value to the attendee should be given for each seminar or congress. This feedback should be taken into account when selecting future seminars.

3.10.   For all managers or personnel in leading positions, seminars on leadership must be available.

3.11.   Seminars on communication and cooperation, environmental protection, industrial safety, and quality are mandatory for all.

3.12.   As the result of the regular performance evaluations or by request, further trainings or seminars may be required.

4. RESPONSIBILITY FOR POLICY IMPLEMENTATION: NAME AND TITLE/FUNCTION OF AUTHORIZED PERSON(S)

5. RULING PERTINENT TO EXISTING PROCEDURES

6. RULING APPLICABLE IN THE CASE OF OUTSOURCING

7. POINTS TO CONSIDER DURING THE DEPLOYMENT OF THIS POLICY

For the application of this and all other policies, general principles apply. (The general principles are given in 7.1–7.7 of P-01.)

7.1.   Carefully monitor the trainings and certification programs organized by professional societies. Regulatory Affairs professional societies offer complete trainings with regard to regional requirements for newcomers. Maintain a list of recommended programs.

7.3.1. The Regulatory Affairs professional must be knowledgeable about the company, its organization, and its medicinal products in order to identify with the goals of Regulatory Affairs.

7.3.2. A senior professional should be appointed as a mentor for each newcomer, because Regulatory Affairs work is usually best acquired by on-the-job training. The mentor should be the primary contact for the newcomer and should suggest and discuss the specific education plan with the newcomer and the department head. The mentor should also monitor the implementation of the education plan and, if required, adjust it. He or she should advise the department head regarding the level of responsibility to be transferred to the newcomer and report progress on a regular basis.

7.3.3. Newcomers should be given regular feedback from the department head, at least after 3, 6, and 12 months. The newcomer, too, should also be encouraged to give feedback. This feedback should result in, if required, corrective actions for this specific education plan and/or future education plans.

7.3.4. A general education plan for similar function(s) should be developed on a generic basis. For each newcomer, however, the education plan should be reviewed with regard to the newcomer's background (e.g., education and previous training) and the function to be performed;

if required, the education plan should be modified as necessary. General education can include seminars on the following:

- Company, organization, products
- Organization and work of departments with which Regulatory Affairs frequently interacts
- Leadership (if leading position)
- Communication and cooperation
- Environmental protection
- Industrial safety
- Quality management

Specialized education can include the following:

- Trainings and certification programs by professional societies
- Extramural seminars and congresses
- In-house training and education

Develop the education plan together with the personnel department. Careful consideration should be given to the exact outline of the seminars, on which a decision will be made whether the seminars will be organized in-house or outsourced. As a rule of thumb, there must be a large group of interested people to make regular in-house seminars profitable. The more specific the group of interested people, the more likely an in-house seminar will make sense.

7.3.5.   Good knowledge, including good interpersonal relationships, of the most important departments with which Regulatory Affairs cooperates is also very important. A presentation of departmental work to Regulatory Affairs professionals helps to ensure good communication.

7.3.6.   The Regulatory Affairs department should consider organizing its own trainings. This would result in an increased feeling of responsibility for quality training within the department in general, a higher degree of education for trainers, and the chance to deliver a very specialized education.

7.3.8.   Trip reports (e.g., congresses) are necessary. Besides spreading the information, they enable the writer to review the results once more.

7.3.9   Bearing in mind the flood of congresses and trainings offered, companies and employees should carefully evaluate the cost/benefit ratio for each educational measure. Feedback from participants helps to evaluate the value of future participation.

7.3.10. As people are the greatest resource you possess in order to attain your goals, seminars on leadership are required.

7.3.11. Specialized education and training (here: in Regulatory Affairs), however, is not sufficient to ensure quality. Therefore, regular trainings should be held on communication and cooperation, environmental protection, industrial safety, and quality.

7.3.12. As result of regular performance evaluations, education on further topics can be offered as necessary.

A Policy of Department XYZ

13: Electronic Submission

Document Type:	Policy
Document Code:	13
	(enter company-specific code)
Title:	Electronic Submission
Date/Revision No.:	DD/MM/YY number xy
Scope:	Global
References:	(enter policies, standards, SOPs of your department/company, or other documents [e.g., guidelines] that should be considered in this context)
	1. Policy on Submission (P-27)
	2. Policy on Dossier (P-11)
	3. Policy on Global Dossier (P-15)
	4. Policy on Tools (P-29)
	5. Policy on Information Technology (P-18)
	6. Departmental operational procedures
Authorization:	
	Signature of authorized person(s)
	Name of authorized person(s)
	Job title/Function of authorized person(s)
Issue Date:	DD/MM/YY
Implementation Date:	DD/MM/YY

1. PURPOSE

This policy covers the work associated with the generation of an electronic submission.

2. DEFINITIONS

The key terms pertaining to this policy should be defined here. As there are no uniform and globally accepted definitions available, please develop your own definitions. In this way, the language of the staff of your organization can be incorporated.

- The term *electronic submission* signifies a submission that contains some or all of its information in an electronic format.

- The term *submission* signifies a country-specific compilation of documents for a specific regulatory purpose (e.g., application for clinical trial authorization or application for marketing authorization) for a developmental or already marketed medicinal product in a structured form according to national regulatory requirements. It is based on the dossier, or, if applicable, the global dossier. It may contain additional national documents (e.g., national leaflets or application forms).

3. STATEMENT OF POLICY

This section covers the actual rulings that should be complied with when working according to the principles of the quality system. For the topic *electronic submission,* rulings should be available for the following items:

3.1.   The electronic submission is based on either the submission (1), the dossier (2), or, if applicable, the global dossier (3).

3.2.   Responsibility for generation and documentation of the electronic submission is with the appointed Regulatory Affairs manager. Assignments are documented in the submission assignments listing (1).

3.3.   The electronic submission is generated after an official decision is made to submit an electronic application for a medicinal product in a specific country. The prerequisites are favourable cost-benefit ratios and agreement with the Regulatory Body.

3.4.   The electronic submission should be generated in a timely fashion. The policies on submission (1), tools (4), and information technology (5) apply.

3.5.   Each department responsible for generating an electronic submission provides a written procedure for processes and responsibilities (6).

4. RESPONSIBILITY FOR POLICY IMPLEMENTATION: NAME AND TITLE/FUNCTION OF AUTHORIZED PERSON(S)

5. RULING PERTINENT TO EXISTING PROCEDURES

6. RULING APPLICABLE IN THE CASE OF OUTSOURCING

7. POINTS TO CONSIDER DURING THE DEPLOYMENT OF THIS POLICY

For the application of this and all other policies, general principles apply. (The general principles are given in 7.1–7.7 of P-01.)

7.2.   Electronic submissions might be paper submissions that contain additional summary parts on disc in Microsoft Word® or WordPerfect®, or a Computer Assisted New Drug Application (CANDA) using SAS® files.

7.3.1. The source data depend on the process. Most companies first complete a paper documentation, then create the electronic form; however, it may be possible in the future to maintain the global dossier electronically and from this create the electronic submission, with the paper format as a printout.

7.3.3. Electronic submissions should be generated only as needed. Though the process is beneficial in itself because it augments the transparency of processes, it is capacity- and time-consuming. Today, there is no routine procedure for both the industry and Regulatory Bodies. Therefore, it is of the utmost importance to determine early in the process not only whether the Regulatory Body is willing to accept an electronic submission and whether there is a real interest by the designated reviewer(s). Closely monitor regulatory requirements. Make sure that in-house decision making is well documented and that you have the full support of upper management.

7.3.5. Develop the SOP(s) on generating an electronic submission.

A Policy of Department XYZ

14: Environmental Protection

Document Type:	Policy
Document Code:	14
	(enter company-specific code)
Title:	Environmental Protection
Date/Revision No.:	DD/MM/YY number xy
Scope:	Global
References:	(enter policies, standards, SOPs of your department/company, or other documents [e.g., guidelines] that should be considered in this context)
Authorization:	
	Signature of authorized person(s)
	Name of authorized person(s)
	Job title/Function of authorized person(s)
Issue Date:	DD/MM/YY
Implementation Date:	DD/MM/YY

1. PURPOSE

This policy is a set of rules developed to specify the principles that apply to environmental protection in the work of Regulatory Affairs worldwide. Within the company, Regulatory Affairs has an obligation to regulate the use of energy and materials with regard to environmental protection. Environmental protection in the field of regulatory affairs is less obvious than in other departments. Unfortunately, it is an area that is all too often neglected. However, for cost-effectiveness, this is an area that must be regulated. Therefore, the objective of this policy is to ensure adequate environmental protection by Regulatory Affairs staff in order to reach and maintain the required high standard of quality in terms of the function within Regulatory Affairs worldwide.

2. DEFINITIONS

The key terms pertaining to this policy should be defined here. As there are no uniform and globally accepted definitions available, please develop your own definitions. In this way, the language of the staff of your organization can be incorporated. The term *environmental protection* in this policy means the responsible use of energy and material within Regulatory Affairs.

3. STATEMENT OF POLICY

This section covers the actual rulings that should be complied with when working according to the principles of the quality system. For the topic *environmental protection,* rulings should be available for the following items:

3.1.   The responsible use of energy

3.2.   The responsible use of use raw materials

3.3.   The avoidance of waste

3.4.   The responsible disposal of waste

4. RESPONSIBILITY FOR POLICY IMPLEMENTATION: NAME AND TITLE/FUNCTION OF AUTHORIZED PERSON(S)

5. RULING PERTINENT TO EXISTING PROCEDURES

6. RULING APPLICABLE IN THE CASE OF OUTSOURCING

7. POINTS TO CONSIDER DURING THE DEPLOYMENT OF THIS POLICY

For the application of this and all other policies, general principles apply. (The general principles are given in 7.1–7.7 of P-01.)

7.1. Explore possibilities for energy saving: use energy-saving bulbs, put machines on standby or turn them off if not used, monitor temperature settings of central heating and air conditioning.

7.2. The responsible use of energy and materials includes the use of water and paper.

7.3. Consider at least the following:

7.3.1. Waste disposal should be considered when purchasing office supplies/equipment. Explore the possibilities of using recycled material (e.g., paper).

7.3.2. In order to avoid extensive paper waste, determine who requires specific documentation. Before sending out documents, be sure that the information is important for the recipient. Attachments should be avoided when possible; a solution might be to indicate that they are on file (i.e., available on demand).

Printouts should be done only when needed. Additionally, limitations can be activated by the number of pages. Corrections in texts from word processing systems should be done on screen rather than on paper. If a printout is required, the back side of former printouts can be used to save paper. Careful planning can help avoid unnecessary copies of documents or dossiers.

E-mail should be used for interoffice mail within the department/company.

7.3.3. When compiling dossiers/submissions, no unnecessary materials should be used.

The quality of presentations is usually increased when fewer transparencies are presented. As rule of thumb, no more than 5–10 transparencies should be required (or used) per presentation.

7.3.4. Explore waste separation and recycling possibilities (e.g., separating paper and other waste; shred confidential documents to allow for recycling).

A Policy of Department XYZ

15: Global Dossier

Document Type:	Policy
Document Code:	15
	(enter company-specific code)
Title:	Global Dossier
Date/Revision No.:	DD/MM/YY number xy
Scope:	Global
References:	(enter policies, standards, SOPs of your department/company, or other documents [e.g., guidelines] that should be considered in this context)
	1. Standard for Global Dossier (S-15.01)
	2. Policy on Documents for Regulatory Purposes (P-10)
	3. Policy on Project Assignments (P-23)
	4. Departmental operational procedures
Authorization:	
	Signature of authorized person(s)
	Name of authorized person(s)
	Job title/Function of authorized person(s)
Issue Date:	DD/MM/YY
Implementation Date:	DD/MM/YY

1. PURPOSE

This policy is a set of rules developed to govern the generation of a global dossier by Regulatory Affairs worldwide.

2. DEFINITIONS

The key terms pertaining to this policy should be defined here. As there are no uniform and globally accepted definitions available, please develop your own definitions. In this way, the language of the staff of your organization can be incorporated.

- The term *dossier* signifies a compilation of documents for a specific regulatory purpose (e.g., application for clinical trial authorization or application for marketing authorization), in a specified country(ies) for a developmental or already marketed medicinal product in a structured form (i.e., submission-like). If applicable, it is a subset of the global dossier. The dossier is the basis for the submission(s).

- The term *global dossier* signifies a compilation of all documents required for international regulatory purpose(s) for a developmental or already marketed medicinal product. It is maintained continuously throughout the life cycle of the medicinal product and serves as a repository for the generation of dossiers and submissions.

- The term *submission* signifies a country-specific compilation of documents for a specific regulatory purpose (e.g., application for clinical trial authorization or application for marketing authorization) for a developmental or already marketed medicinal product in a structured form according to national regulatory requirements. It is based on the dossier, or, if applicable, the global dossier. It may contain additional national documents (e.g., national leaflets or application forms).

3. STATEMENT OF POLICY

This section covers the actual rulings that should be complied with when working according to the principles of the quality system. For the topic *global dossier*, rulings should be available for the following items:

3.1.   For each developmental or already marketed medicinal product, a global dossier is generated and maintained throughout the entire life cycle of the medicinal product. The structure of the global dossier is defined (1). The format and content of elements are defined by internal company standards (2).

3.2.   Responsibility for generation and maintenance of the global dossier is with the appointed Regulatory Affairs manager. Assignments are documented in the listing of project assignments (3).

3.3.   The global dossier is prepared after an official company decision is made to develop a medicinal product. It is maintained throughout the entire life cycle of the medicinal product. Updates are processed in a timely fashion after availability at Regulatory Affairs. Changes to the global dossier are adequately documented.

3.4.   Each department responsible for the generation and maintenance of a global dossier provides a written procedure for processes and responsibilities (4).

4.   RESPONSIBILITY FOR POLICY IMPLEMENTATION: NAME AND TITLE/FUNCTION OF AUTHORIZED PERSON(S)

5.   RULING PERTINENT TO EXISTING PROCEDURES

6.   RULING APPLICABLE IN THE CASE OF OUTSOURCING

7.   POINTS TO CONSIDER DURING THE DEPLOYMENT OF THIS POLICY

For the application of this and all other policies, general principles apply. (The general principles are given in 7.1–7.7 of P-01.)

7.1.   One of the most important Regulatory Affairs functions is to generate dossiers for regulatory purposes (e.g., to obtain a clinical trial authorization, a marketing authorization, or a renewal of a marketing authorization). The technical part of this process is reflected by the policies on global dossier, dossier, and submission. For companies operating in a single market, the policies for dossiers and submissions should be combined.

7.2.   For companies continuously maintaining a global dossier, it may be identical with the dossier. However, most companies provide national Regulatory Affairs units only with the relevant subset for the intended submission, not the entire documentation. Usually, the dossier is provided in a submission-like structure to facilitate the generation of the submission. Based on your organization needs, define whether the dossier will or will not be paginated and cross-referenced.

7.3.1. The standard for the generation of the global dossier should be adapted to the needs of your organization.

7.3.3. Define the time frame and adequate documentation requirements for your organization.

7.3.4. Develop the SOP(s) for generating a global dossier.

A Policy of Department XYZ

16: Import/Export

Document Type:	Policy
Document Code:	16
	(enter company-specific code)
Title:	Import/Export
Date/Revision No.:	DD/MM/YY number xy
Scope:	Global
References:	(enter policies, standards, SOPs of your department/company, or other documents [e.g., guidelines] that should be considered in this context)
	1. Policy on Contact with Regulatory Body (P-08)
	2. Departmental operational procedures
	3. List of responsible persons for import/export by country
Authorization:	
	Signature of authorized person(s)
	Name of authorized person(s)
	Job title/Function of authorized person(s)
Issue Date:	DD/MM/YY
Implementation Date:	DD/MM/YY

1. PURPOSE

This policy is a set of rules developed to govern the import/export of the company's substances and medicinal products for Regulatory Affairs world-wide.

2. DEFINITIONS

The key terms pertaining to this policy should be defined here. As there are no uniform and globally accepted definitions available, please develop your own definitions. In this way, the language of the staff of your organization can be incorporated. The term *substance* or *medicinal product* in this policy means both development/research products and already marketed medicinal products for which the company has legal responsibility, be it as a contract manufacturer, the person responsible for bringing the product into the market, the partner in co-marketing, the receiver or giver of a license, the Drug Master File (DMF) holder or letter of authorization holder, or the partner in joint venture development.

3. STATEMENT OF POLICY

This section covers the actual rulings that should be complied with when working according to the principles of the quality system. For the topic *import/export,* rulings should be available for the following items:

3.1.   Define responsibility for import/export according to your organization and procedures.

3.2.   Regulatory Affairs is responsible for contact or activities with the Regulatory Bodies in regard to import/export. The policy on contact with regulatory body (1) applies.

3.3.   The person(s) responsible for import/export should follow the appropriate organizational procedure(s) (2) to meet the respective national regulatory requirements.

3.4.   A list of responsible persons for import/export by country will be generated and maintained (3).

**4.   RESPONSIBILITY FOR POLICY IMPLEMENTATION:
     NAME AND TITLE/FUNCTION OF AUTHORIZED PERSON(S)**

5.   RULING PERTINENT TO EXISTING PROCEDURES

6.   RULING APPLICABLE IN THE CASE OF OUTSOURCING

**7.   POINTS TO CONSIDER
     DURING THE DEPLOYMENT OF THIS POLICY**

For the application of this and all other policies, general principles apply. (The general principles are given in 7.1–7.7 of P-01.)

7.3.3. Regulatory Affairs is responsible for the monitoring of the regulatory environment; in this case, the applicable regulations concerning import/export. If Regulatory Affairs is responsible for import/export, develop and add the standard and/or SOP to the Quality Manual.

7.3.4. Develop the list of responsible persons according to the procedures of your organization.

A Policy of Department XYZ

17: Information Management

Document Type:	Policy
Document Code:	17
	(enter company-specific code)
Title:	Information Management
Date/Revision No.:	DD/MM/YY number xy
Scope:	Global
References:	(enter policies, standards, SOPs of your department/company, or other documents [e.g., guidelines] that should be considered in this context)
	1. Policy on Information Technology (P-18)
	2. Policy on Crisis Management (P-09)
	3. Policy on Change Alert/Authorization Process (P-06)
	4. Policy on Promotion/Advertising Compliance (P-24)
	5. Policy on Contact Report (P-07)
	6. Policy on Periodic Safety Update Report (P-22)
	7. Policy on Labeling (P-20)
	8. Policy on Contact with Regulatory Body (P-08)
	9. Policy on Education/Training (P-12)

Authorization:

_____
Signature of authorized person(s)

_____
Name of authorized person(s)

_____
Job title/Function of authorized person(s)

Issue Date: DD/MM/YY
_____

Implementation Date: DD/MM/YY
_____

1. PURPOSE

This policy is a set of rules developed to specify the principles that apply to the management of information on the company's substances and medicinal products by Regulatory Affairs worldwide with the objective of

- Ensuring compliance with legal and/or business obligations by the adequate transfer of information to the appropriate contact partners

- Avoiding delays in information, misrepresentation, and lack of information

- Assuring correct interaction with other procedures/departments involved within Regulatory Affairs and its contact partners worldwide

2. DEFINITIONS

The key terms pertaining to this policy should be defined here. As there are no uniform and globally accepted definitions available, please develop your own definitions. In this way, the language of the staff of your organization can be incorporated.

- The term *substance* or *medicinal product* in this policy means both development/research products and already marketed medicinal products for which the company has legal responsibility, be it as a contract manufacturer, the person responsible for bringing the product into the market, the partner in co-marketing, the receiver or giver of a license, the Drug Master File (DMF) holder or letter of authorization holder, or the partner in joint venture development.

- The term *information* in this context means any knowledge on the company's substances and/or medicinal products that might be relevant for partners within Regulatory Affairs or other contact partners as defined by legal and/or business obligations. This covers information received directly, by phone, E-mail, fax, letter, or other route.

- The term *contact partner* in this context means the person(s), department(s), company(ies), and/or Regulatory Body(ie)s with which Regulatory Affairs personnel interact in a business environment.

- The term *contact with Regulatory Bodies* signifies any communication between Regulatory Affairs and a Regulatory Body directly, by phone, E-mail, fax, telex, or letter, or during a personal meeting.

3.   STATEMENT OF POLICY

This section covers the actual rulings that should be complied with when working according to the principles of the quality system. For the topic *information management,* rulings should be available for the following items:

3.1.   Distribution of information on the company's substances/medicinal products.

3.2.   Requirement to meet applicable laws and regulations and/or business obligations.

3.3.   Communication of information that may substantially impact the legal and/or business obligations of the company without undue delay.

3.4.   Identification and appropriate communication on information that is vital in nature.

3.5.   Truth, plausibility, and reliability of information communicated by Regulatory Affairs.

3.6.   Communication of information that could be of importance to the company.

3.7.   Requirement to meet internal and external customers' needs in terms of level of detail and access time.

3.8.   Requirement to meet general business rules concerning secrecy agreements, rules for noncompany personnel, rules for correspondence and signatures, e.g., such as the following:

- Corporate design/forms for correspondence

- Business correspondence—external

- Business correspondence—internal

- Electronic mail (telex, fax, E-mail) (1)

- Going through the mail

- Secrecy, confidentiality of documents

- Taking confidential business papers beyond company's premises

- Distribution of circular letters, or minutes of regularly recurring meetings

3.9.   Release procedures for lectures and publications.

3.10.   Consult the following policies for more information: crisis management (2), change alert/authorization (3), promotion/advertising compliance (4), contact report (5), periodic Safety Update Report (6), labeling (7).

3.11.   Procedures are required for communication of information on actual or suspected harm or risk connected with the company's substances/medicinal products, or information on other substances/medicinal products that could also apply to the company's substances/medicinal products (regardless whether the information refers to approved or unapproved use of the substances/medicinal products).

3.12.   Procedures in case of suspected illegal or criminal exploitation of the company's work by other companies (e.g., patent infringement and counterfeit products).

3.13.   Procedures on use of insider knowledge.

3.14.   Contact with Regulatory Bodies (8).

3.15.   Procedures for suggestions for improvement should be written down.

3.16.   Regulatory Affairs employees will receive the information they require for their everyday work in adequate level of detail in a timely fashion.

3.17.   Requirement for clear organization and function(s).

3.18.   Regulatory Affairs employees will be given the opportunity to take courses and attend congresses on a need-to-know basis (9).

3.19.   Regulations on subscriptions to newspapers, journals, and other publications.

3.20.   Regulations on the ordering of copies of publications.

3.21.   Procedure for the announcement of visitors and the generation of their agenda.

4. RESPONSIBILITY FOR POLICY IMPLEMENTATION: NAME AND TITLE/FUNCTION OF AUTHORIZED PERSON(S)

5. RULING PERTINENT TO EXISTING PROCEDURES

6. RULING APPLICABLE IN THE CASE OF OUTSOURCING

7. POINTS TO CONSIDER DURING THE DEPLOYMENT OF THIS POLICY

For the application of this and all other policies, general principles apply. (The general principles are given in 7.1–7.7 of P-01.)

7.3.1. Name the responsible department for the distribution of information on the company's substances/medicinal products.

7.3.6. A lack of information is serious mismanagement; therefore, great care should be taken to avoid this.

7.3.8. General rules may seem so evident as to make a policy on secrecy agreements, confidentiality, and so on superfluous; however, practice shows that there should be established standards for easy reference to guarantee quality. A simple thing, such as an incorrect signature on an important communication or agreement, may lead to great damage to the department and/or company.

A Policy of Department XYZ

18: Information Technology

Document Type:	Policy
Document Code:	18
	(enter company-specific code)
Title:	Information Technology
Date/Revision No.:	DD/MM/YY number xy
Scope:	Global
References:	(enter policies, standards, SOPs of your department/company, or other documents [e.g., guidelines] that should be considered in this context)
Authorization:	
	Signature of authorized person(s)
	Name of authorized person(s)
	Job title/Function of authorized person(s)
Issue Date:	DD/MM/YY
Implementation Date:	DD/MM/YY

1. PURPOSE

This policy is a set of rules developed to specify the principles that apply to electronic data processing (EDP) and telecommunication for Regulatory Affairs worldwide with the objective of

- Ensuring compliance with legal and/or business obligations

- Avoiding delays and/or costs because of incompatible or otherwise unsuitable hardware and/or software

- Assuring correct interaction with other procedures/departments involved within Regulatory Affairs and its contact partners worldwide

2. DEFINITIONS

The key terms pertaining to this policy should be defined here. As there are no uniform and globally accepted definitions available, please develop your own definitions. In this way, the language of the staff of your organization can be incorporated.

- The term *information* in this context means any knowledge on the company's substances and/or medicinal products that might be relevant for partners within Regulatory Affairs or other contact partners as defined by legal and/or business obligations. This covers information received directly, by phone, E-mail, fax, letter, or other route.

- The term *contact partner* in this context means the person(s), department(s), company(ies), and/or Regulatory Body(ies) with which Regulatory Affairs personnel interacts in a business environment.

3. STATEMENT OF POLICY

This section covers the actual rulings that should be complied with when working according to the principles of the quality system. For the topic *information technology,* rulings should be available for the following items:

3.1.   To ensure data security, every user must have a unique user ID and password(s). Access to personal computers and electronic networks should be regulated via a security check. The user ID and password(s) must be kept confidential and must not be used by other persons. The user ID and password(s) must be assigned by appropriately defined and validated procedures. These guidelines also apply when replacement workers stand in for someone who is ill or on vacation.

3.2. Legal and/or business obligations will be appropriately met. The rights of the individual will be respected, regardless of whether the employee is in Regulatory Affairs or another department.

3.3. In order to ensure the cost-effective use of personal computers (PCs) as well as compatibility and integratability into existing networks within the company, the acquisition of such hardware and software should be according to standards developed by the responsible person/department (usually the Information Technology department). This also applies to portable PCs.

3.4. Users of PCs with external connections must take appropriate safety precautions against hackers and viruses.

3.5. There is no guarantee on availability and/or reliability of information or services on the Internet and no guarantee of confidentiality or integrity of information transfer. Therefore, the Internet should be used only on a need-to-know basis, with adequate measures for protection (e.g., with a fire wall).

3.6. In order to ensure the cost-effective use of telecommunication equipment as well as compatibility and integratability into existing communication networks within the company, the acquisition of such equipment should be according to standards developed by the responsible person/department (usually the Information Technology department). This also applies for to digital equipment (e.g., ISDN telephones, ISDN PC cards, and cellular telephones).

4. RESPONSIBILITY FOR POLICY IMPLEMENTATION: NAME AND TITLE/FUNCTION OF AUTHORIZED PERSON(S)

5. RULING PERTINENT TO EXISTING PROCEDURES

6. RULING APPLICABLE IN THE CASE OF OUTSOURCING

7. POINTS TO CONSIDER DURING THE DEPLOYMENT OF THIS POLICY

For the application of this and all other policies, general principles apply. (The general principles are given in 7.1–7.7 of P-01.)

Regulatory Affairs professional(s) should not attempt to finalize or implement this policy without help from Information Technology professionals (internal and/or external)! In the context of this policy, you should think about who are the contact partners and what the optimal communication would look like. The draft policy deals primarily with avoiding the pitfalls; however, you may also wish to think about ways to improve existing communication.

7.1.   The aspects to keep in mind concerning EDP and telecommunication are legal/business obligations, compatibility, and communication. Legal obligations might include the following:

- Protection of the rights of individuals concerning the electronic storage of personal information

- Restrictions concerning the storage of data that could be used to control employee performance.

Business obligations might be the confidentiality of certain data.

Compatibility is a key issue if several individuals work and communicate using EDP and/or telecommunication equipment. Compatible systems help to save time, costs, and capacity.

The optimal transfer of information is vital for good communication.

7.3.1.   Every user should assume responsibility. Therefore, unique user ID and password(s) for each user are a must. You may wish to choose computer-assisted generation of password(s) where random password(s) are suggested at regular intervals. Experience shows that creativity is frequently wanting when it comes to the selection of passwords. Because of the danger of unauthorized use, password(s) must be changed on a regular basis. It is NOT acceptable for a replacement worker to use the same password(s) as the person for whom he or she is replacing. In these specified cases, access rights should be granted instead. It is vital that appropriately defined and validated procedures cover all aspects of user ID and passwords.

7.3.2.   The rights of the individual should be respected. Especially with electronic databases, great care must be taken so that information on individuals does not violate their rights.

7.3.3.   Adequate electronic tools are of key importance to process quality. This does not necessarily mean the most up-to-date hardware and software. However, development in this sector should be carefully screened and equipment provided to users based on a benefit/cost ratio. Also monitor the situation at contact partners with whom you wish to interact (e.g., Regulatory Bodies). This applies especially to word processing programs and data exchange formats.

7.3.4.   External communication should be allowed only on a need-basis as it involves the risk of a breach of confidentiality. Up-to-date measures against hackers and viruses should be taken. It is recommended to analyze the possible dangers of each type of external communication and to make adequate provisions. Commercial virus protection services should be evaluated.

7.3.5.   The Internet is of special interest to the Regulatory Affairs professional, especially as it offers up-to-date information on Regulatory Bodies and guidelines issued by the European Agency for the Evaluation of Medicinal Products (EMEA) and the Food and Drug Administration (FDA). However, it must be kept in mind that the information presented may not always be complete and/or reliable.

7.3.6.   Up-to-date equipment can prove to be a major benefit. Consider installing answering machines and voice mail.

A Policy of Department XYZ

19: Inspection

Document Type:	Policy
Document Code:	19
	(enter company-specific code)
Title:	Inspection
Date/Revision No.:	DD/MM/YY number xy
Scope:	Global
References:	(enter policies, standards, SOPs of your department/company, or other documents [e.g., guidelines] that should be considered in this context)
	1. Policy on Contact with Regulatory Body (P-08)
	2. Policy on Contact Report (P-07)
	3. Policy on Archiving Management (P-05)
Authorization:	
	Signature of authorized person(s)
	Name of authorized person(s)
	Job title/Function of authorized person(s)
Issue Date:	DD/MM/YY
Implementation Date:	DD/MM/YY

1. PURPOSE

This policy is a set of rules developed to govern inspections of a company by Regulatory Bodies worldwide.

2. DEFINITIONS

The key terms pertaining to this policy should be defined here. As there are no uniform and globally accepted definitions available, please develop your own definitions. In this way, the language of the staff of your organization can be incorporated.

- The term *Regulatory Body* in the context of this policy will mean any Regulatory Body authorized to carry out inspections of the company.

- The term *inspection* in the context of this policy signifies inspections by Regulatory Bodies with regard to Regulatory Affairs (e.g., inspection before issuing a marketing authorization).

3. STATEMENT OF POLICY

This section covers the actual rulings that should be complied with when working according to the principles of the quality system. For the topic *inspection,* rulings should be available for the following items:

3.1.   Regulatory Affairs is responsible for the organization of such inspections and accompanies all inspectors. The policy for contact with Regulatory Bodies applies (1).

3.2.   If the inspection is preannounced, Regulatory Affairs is responsible for adequate preparation (e.g., internal audit, preparation for probable questions, strategy) with the departments/scientific disciplines concerned.

3.3.   In addition to the Regulatory Body's inspection report, Regulatory Affairs will write a contact report (2). The inspection and contact reports will be adequately archived (3).

3.4.   If required, areas for improvement will be identified and adequate measures suggested, and implementation will be monitored by Regulatory Affairs. If required, an internal postinspection meeting will be held with the departments/scientific disciplines inspected, with the aim of defining measures for improvement. The organization of such a meeting is the responsibility of Regulatory Affairs.

4. **RESPONSIBILITY FOR POLICY IMPLEMENTATION:
NAME AND TITLE/FUNCTION OF AUTHORIZED PERSON(S)**

5. **RULING PERTINENT TO EXISTING PROCEDURES**

6. **RULING APPLICABLE IN THE CASE OF OUTSOURCING**

7. **POINTS TO CONSIDER
DURING THE DEPLOYMENT OF THIS POLICY**

For the application of this and all other policies, general principles apply. (The general principles are given in 7.1–7.7 of P-01.)

7.3.1. The organization/accompanying of inspections is not always the re-sponsibility of Regulatory Affairs. However Regulatory Affairs is best suited for this task because of their close contact with Regulatory Bodies and intimate knowledge of their procedures.

A Policy of Department XYZ

20: Labeling

Document Type:	Policy
Document Code:	20
	(enter company-specific code)
Title:	Labeling
Date/Revision No.:	DD/MM/YY number xy
Scope:	Global
References:	(enter policies, standards, SOPs of your department/company, or other documents [e.g., guidelines] that should be considered in this context)
	1. Standard for Labeling (S-20.01)
	2. Policy on Promotion/Advertising Compliance (P-24)
	3. Policy on Change Alert/Authorization Process (P-06)
Authorization:	
	Signature of authorized person(s)
	Name of authorized person(s)
	Job title/Function of authorized person(s)
Issue Date:	DD/MM/YY
Implementation Date:	DD/MM/YY

1.   PURPOSE

This policy is a set of rules developed to govern the generation and maintenance of labeling with the purpose of specifying and harmonizing essential statements on the company's medicinal products worldwide.

2.   DEFINITIONS

Here the key terms pertaining to this policy should be defined. As there are no uniform and globally accepted definitions available, please develop your own definitions. In this way, the language of the staff of your organization can be incorporated. The term *labeling* means a document reflecting the company's actual state of knowledge on its medicinal products, by documenting per medicinal product the scientifically relevant and essential statements. This labeling is the basis for patient leaflet(s), professional information(s), and promotional material(s).

3.   STATEMENT OF POLICY

This section covers the actual rulings that should be complied with when working according to the principles of the quality system. For the topic *labeling*, rulings should be available for the following items:

3.1.   Labeling shows the information per medicinal product that the company requires to be reflected in national labeling and promotional material. The general content is defined by a standard (1).

3.2.   Labeling is an element of the dossier and, if applicable, the global dossier. However, it is for in-house use only. It is the basis for national labeling and, if applicable, promotional material.

3.3.   The responsibility for the generation and maintenance of labeling is with Regulatory Affairs, as well as the responsibility for national labeling. The policy on promotion and advertising also applies (2).

3.4.   For variations to labeling, the change alert/authorization process applies (3).

4. RESPONSIBILITY FOR POLICY IMPLEMENTATION: NAME AND TITLE/FUNCTION OF AUTHORIZED PERSON(S)

5. RULING PERTINENT TO EXISTING PROCEDURES

6. RULING APPLICABLE IN THE CASE OF OUTSOURCING

7. POINTS TO CONSIDER DURING THE DEPLOYMENT OF THIS POLICY

For the application of this and all other policies, general principles apply. (The general principles are given in 7.1–7.7 of P-01.)

7.3.2. Elaborate on the suggested standard for labeling as required by your organization and the actual regulatory environment.

7.3.3. Some companies require translations of the exact wording of in-house labeling to be used for national labeling; others require only the reflection of core statements. It is the company's responsibility to supply adequate information to affiliates, Regulatory Bodies, and, ultimately, to patients, and thus to ensure correct labeling. Beware of very marketing-oriented texts. On the other hand, do not hesitate to discuss critical issues with the Regulatory Bodies.

A Policy of Department XYZ

21: Outsourcing

Document Type:	Policy
Document Code:	21
	(enter company-specific code)
Title:	Outsourcing
Date/Revision No.:	DD/MM/YY number xy
Scope:	Global
References:	(enter policies, standards, SOPs of your department/company, or other documents [e.g., guidelines] that should be considered in this context)
	1. Policy on Crisis Management (P-09)
Authorization:	
	Signature of authorized person(s)
	Name of authorized person(s)
	Job title/Function of authorized person(s)
Issue Date:	DD/MM/YY
Implementation Date:	DD/MM/YY

1. PURPOSE

This policy is a set of rules developed to govern the outsourcing of Regulatory Affairs work.

2. DEFINITIONS

The key terms pertaining to this policy should be defined here. As there are no uniform and globally accepted definitions available, please develop your own definitions. In this way, the language of the staff of your organization can be incorporated.

- The term *outsourcing* for the purpose of this policy means in-sourcing, co-sourcing, or outsourcing of Regulatory Affairs work to external parties.

- The term *external parties* for the context of this policy means individuals, consultants, or Contract Research Organizations (CROs) outside Regulatory Affairs that can supply functions required for Regulatory Affairs.

3. STATEMENT OF POLICY

This section covers the actual rulings that should be complied with when working according to the principles of the quality system. For the topic *outsourcing,* rulings should be available for the following items:

3.1.   If required, Regulatory Affairs work may be outsourced to external parties.

3.2.   The criteria for the selection of such external parties will be adequate quality and cost-effectiveness. If available, previous experience will also be considered (see under section 3.5. of this policy).

3.3.   The person/department outsourcing will be responsible for the conduct and performance of such external parties and will take the necessary measures to ensure that they appropriately meet all business and legal obligations on behalf of such person/department.

3.4.   The person/department outsourcing will make sure that functions performed by such third parties on behalf of Regulatory Affairs meet the standards established by the quality system.

3.5.   A list will be developed and maintained on such external parties that might stand in for Regulatory Affairs in cases of crisis (1).

3.6.   The actual performance of such external parties for Regulatory Affairs will be evaluated and experience documented in the list (see under section 3.5. of this policy).

3.7.   Special rulings applicable in the case of outsourcing will be defined in the individual policies of the quality system.

4.   RESPONSIBILITY FOR POLICY IMPLEMENTATION: NAME AND TITLE/FUNCTION OF AUTHORIZED PERSON(S)

5.   RULING PERTINENT TO EXISTING PROCEDURES

6.   RULING APPLICABLE IN THE CASE OF OUTSOURCING

7.   POINTS TO CONSIDER DURING THE DEPLOYMENT OF THIS POLICY

For the application of this and all other policies, general principles apply. (The general principles are given in 7.1–7.7 of P-01.)

7.3.1.   This should not be interpreted as a call for the outsourcing of some or all of Regulatory Affairs' work; however, there must be provisions in place in case some work must be outsourced in order to achieve the goals of Regulatory Affairs.

7.6.   This is not a typing error: Even outsourcing can be outsourced; namely, you can delegate the responsibility of selecting and monitoring CROs to someone else. In this case, you need to make the applicable provisions.

A Policy of Department XYZ

22: Periodic Safety Update Report

Document Type:	Policy
Document Code:	22
	(enter company-specific code)
Title:	Periodic Safety Update Report
Date/Revision No.:	DD/MM/YY number xy
Scope:	Global
References:	(enter policies, standards, SOPs of your department/company, or other documents [e.g., guidelines] that should be considered in this context)
	1. Standard for Periodic Safety Update Report (to be developed by department)
	2. Departmental operational procedures
Authorization:	
	Signature of authorized person(s)
	Name of authorized person(s)
	Job title/Function of authorized person(s)
Issue Date:	DD/MM/YY
Implementation Date:	DD/MM/YY

1.   PURPOSE

This policy is a set of rules developed to govern the generation of periodic Safety Update Reports by Drug Safety and Regulatory Affairs worldwide.

2.   DEFINITIONS

The key terms pertaining to this policy should be defined here. As there are no uniform and globally accepted definitions available, please develop your own definitions. In this way, the language of the staff of your organization can be incorporated. The term *periodic Safety Update Report* signifies a document for regulatory purposes issued by Drug Safety and Regulatory Affairs for the company's marketed medicinal products for safety updates or pharmacovigilance according to regulatory requirements. It is part of, if applicable, the global dossier, and, if nationally required, part of dossier(s) and submission(s).

3.   STATEMENT OF POLICY

This section covers the actual rulings that should be complied with when working according to the principles of the quality system. For the topic *periodic Safety Update Report,* rulings should be available for the following items:

3.1.   For each of the company's marketed medicinal products, the periodic Safety Update Report will be generated and updated according to current regulations.

3.2.   The responsibility for the periodic Safety Update Report lies with Drug Safety and Regulatory Affairs.

3.3.   Develop a standard for the structure, format, and content of the periodic Safety Update Report (1). Develop procedures for all departments responsible for generating the periodic Safety Update Report (2).

4.   RESPONSIBILITY FOR POLICY IMPLEMENTATION: NAME AND TITLE/FUNCTION OF AUTHORIZED PERSON(S)

5.   RULING PERTINENT TO EXISTING PROCEDURES

6.   RULING APPLICABLE IN THE CASE OF OUTSOURCING

7.   POINTS TO CONSIDER DURING THE DEPLOYMENT OF THIS POLICY

For the application of this and all other policies, general principles apply. (The general principles are given in 7.1–7.7 of P-01.)

7.3.1.   Regulatory Affairs is responsible for closely monitoring the regulatory environment.

7.3.2.   Principal ownership and responsibility lies with Drug Safety; however, Regulatory Affairs shares the responsibility. The periodic Safety Update Report is a key document for regulatory purposes to which Regulatory Affairs also contributes, for example, the marketing history. Regulatory Affairs is the internal customer of the periodic Safety Update Report.

7.3.3.   Develop the standard and the respective SOPs.

A Policy of Department XYZ

23: Project Assignments

Document Type:	Policy
Document Code:	23
	(enter company-specific code)
Title:	Project Assignments
Date/Revision No.:	DD/MM/YY number xy
Scope:	Global
References:	(enter policies, standards, SOPs of your department/company, or other documents [e.g., guidelines] that should be considered in this context)
	1. Policy on Global Dossier (P-15)
	2. Standard for Project Assignments Listing (to be developed by department)
Authorization:	
	Signature of authorized person(s)
	Name of authorized person(s)
	Job title/Function of authorized person(s)
Issue Date:	DD/MM/YY
Implementation Date:	DD/MM/YY

1. PURPOSE

This policy is a set of rules developed to govern project assignments and to provide for a listing of these.

2. DEFINITIONS

The key terms pertaining to this policy should be defined here. As there are no uniform and globally accepted definitions available, please develop your own definitions. In this way, the language of the staff of your organization can be incorporated. The term *global dossier* signifies a compilation of all documents required for international regulatory purpose(s) for a developmental or already marketed medicinal product. It is maintained continuously throughout the life cycle of the medicinal product and serves as a repository for the generation of dossiers and submissions.

3. STATEMENT OF POLICY

This section covers the actual rulings that should be complied with when working according to the principles of the quality system. For the topic *project assignments,* rulings should be available for the following items:

3.1.   The responsibility for the generation and maintenance of the global dossier (1) for the company's developmental or marketed medicinal products is with the appointed Regulatory Affairs manager.

3.2.   Regulatory Affairs will generate and maintain a listing of responsible Regulatory Affairs managers per medicinal product. Distribution will be according to an agreed on standard distribution list of interested parties. Every update/modification will be made available to such interested parties within five working days (2).

**4.   RESPONSIBILITY FOR POLICY IMPLEMENTATION:
NAME AND TITLE/FUNCTION OF AUTHORIZED PERSON(S)**

5.   RULING PERTINENT TO EXISTING PROCEDURES

6.   RULING APPLICABLE IN THE CASE OF OUTSOURCING

**7.   POINTS TO CONSIDER
DURING THE DEPLOYMENT OF THIS POLICY**

For the application of this and all other policies, general principles apply. (The general principles are given in 7.1–7.7 of P-01.)

7.3.1. Make it clear in your organization that the responsibility for the content of the individual documents for regulatory purposes incorporated into the global dossier remains with the respective authors/ scientific disciplines.

7.3.2. Consider making the list electronically available.

A Policy of Department XYZ

24: Promotion/Advertising Compliance

Document Type:	Policy
Document Code:	24
	(enter company-specific code)
Title:	Promotion/Advertising Compliance
Date/Revision No.:	DD/MM/YY number xy
Scope:	Global
References:	(enter policies, standards, SOPs of your department/company, or other documents [e.g., guidelines] that should be considered in this context)
	1. Policy on Labeling (P-20)
	2. Policy on Information Management (P-17)
	3. Policy on Change Alert/Authorization Process (P-06)
	4. Policy on Submission (P-27)
	5. Policy on Archiving Management (P-05)
Authorization:	
	Signature of authorized person(s)
	Name of authorized person(s)
	Job title/Function of authorized person(s)
Issue Date:	DD/MM/YY
Implementation Date:	DD/MM/YY

1. PURPOSE

This policy is a set of rules developed to specify the procedures by which promotion/advertising compliance is assured by Regulatory Affairs worldwide.

2. DEFINITIONS

The key terms pertaining to this policy should be defined here. As there are no uniform and globally accepted definitions available, please develop your own definitions. In this way, the language of the staff of your organization can be incorporated. The term *promotion/advertising* in the context of this policy refers to any published information on the company's substances or medicinal products that is published with a view of making the product known and/or augmenting sales (e.g., newspaper ads, television commercials). It excludes patient information and physician's information, which is regulated by the policy on labeling (1).

3. STATEMENT OF POLICY

This section covers the actual rulings that should be complied with when working according to the principles of the quality system. For the topic *promotion/advertising compliance,* rulings should be available for the following items:

3.1.   Responsibility for promotion/advertising compliance is with Regulatory Affairs. (In case responsibility is with another functional unit, ensure that promotion/advertising complies with approved labeling by involving Regulatory Affairs in the approval process.)

3.2.   Promotion/advertising must never be unethical.

3.3.   Promotion/advertising must comply in a timely fashion with legal and/or business obligations, and the policies on labeling (1) and information management (2).

3.4.   Adequate procedures for compliance and release will be established, including the documentation of changes; change management applies (3).

3.5.   The policies for submission (4) and archiving management (5) apply.

4.   RESPONSIBILITY FOR POLICY IMPLEMENTATION: NAME AND TITLE/FUNCTION OF AUTHORIZED PERSON(S)

5.   RULING PERTINENT TO EXISTING PROCEDURES

6.   RULING APPLICABLE IN THE CASE OF OUTSOURCING

7.   POINTS TO CONSIDER DURING THE DEPLOYMENT OF THIS POLICY

For the application of this and all other policies, general principles apply. (The general principles are given in 7.1–7.7 of P-01.)

7.3.1.   Promotion/advertising is considered to be the responsibility of Regulatory Affairs in some but not all countries. If preapproval is required by Regulatory Bodies for promotional material, Regulatory Affairs is then responsible. For the sake of harmonization of promotional material with registered labeling, it seems advisable to unite the responsibility for both in one function.

7.3.2.   For example, it would be unethical to arouse a fear of illness for promotional purposes.

A Policy of Department XYZ

25: Regulations and Guidelines

Document Type:	Policy
Document Code:	25
	(enter company-specific code)
Title:	Regulations and Guidelines
Date/Revision No.:	DD/MM/YY number xy
Scope:	Global
References:	(enter policies, standards, SOPs of your department/company, or other documents [e.g., guidelines] that should be considered in this context)
	1. Policy on Documents for Regulatory Purposes (P-10)
	2. Policy on Archiving Management (P-05)
	3. Standard for Company Position (to be developed by department)
Authorization:	
	Signature of authorized person(s)
	Name of authorized person(s)
	Job title/Function of authorized person(s)
Issue Date:	DD/MM/YY
Implementation Date:	DD/MM/YY

1. PURPOSE

This policy is a set of rules developed to specify the principles governing for Regulatory Affairs worldwide regarding the collection, evaluation, archiving, and distribution of any relevant guidelines for the company's substances and medicinal products, with the aim of achieving and maintaining a high degree of understanding and knowledge of the regulatory requirements within the organization and to agree on a company position, which provides a basis for the standardization of documents for regulatory purposes (1).

2. DEFINITIONS

The key terms pertaining to this policy should be defined here. As there are no uniform and globally accepted definitions available, please develop your own definitions. In this way, the language of the staff of your organization can be incorporated.

- The term *guidelines* in the context of this policy signifies worldwide guidelines, regulations, laws, relevant publications, position papers, and company experience (e.g., contact reports, deficiency letters, Regulatory Affairs know-how) that may impact the company's substances and/or medicinal products (e.g., medicinal product development, marketing authorization, and surveillance programs).

- The term *company position* in the context of this policy signifies for each important guideline the result of interpretation/discussion by Regulatory Affairs, and, if applicable, other concerned departments/disciplines. It will contain a summary and an evaluation of, for example, critical issues, possible consequences for the company, and recommendations.

- The term *company comment* in the context of this policy signifies a document communicating to Regulatory Bodies comments suggestions for changing a guideline. Typically, it will be in letter format and contain comments, suggestions, and critical issues, with references to the original document. It may be submitted via industry associations and/or Regulatory Affairs professional societies or directly to the Regulatory Bodies.

3. STATEMENT OF POLICY

This section covers the actual rulings that should be complied with when working according to the principles of the quality system. For the topic *regulations and guidelines,* rulings should be available for the following items:

3.1. In order to make guidelines easily accessible, Regulatory Affairs maintains a guideline archive (for regulatory documents, see the policy on archiving management [2]). Regulatory Affairs will be responsible for acquiring copies of relevant new guidelines in a timely fashion (if required with translation) and for making them available, preferably in electronic form. Interested parties will be informed appropriately on a quarterly basis on new guidelines; very important guidelines will be communicated not later than one month after receipt.

3.2. The evaluation/assessment of important guidelines will be coordinated by Regulatory Affairs in a timely fashion. Discussions will be between Regulatory Affairs and the departments/scientific disciplines concerned. The result will be a company position summarized by Regulatory Affairs according to the standard for format and content (3). The company position will made accessible by Regulatory Affairs with cross-references to the original document, preferably in electronic form. Interested parties will be informed appropriately on a quarterly basis.

3.3. If required, a company comment based on the company position will be sent to industry associations, Regulatory Affairs professional societies, and/or Regulatory Bodies. It is the responsibility of Regulatory Affairs to compile the company comment in conjunction with the other departments/scientific disciplines concerned.

3.4. If required, Regulatory Affairs will participate in working parties of industry associations and/or Regulatory Affairs professional societies to introduce the company's position at an early stage of discussion.

4. RESPONSIBILITY FOR POLICY IMPLEMENTATION: NAME AND TITLE/FUNCTION OF AUTHORIZED PERSON(S)

5. RULING PERTINENT TO EXISTING PROCEDURES

6. RULING APPLICABLE IN THE CASE OF OUTSOURCING

7. POINTS TO CONSIDER DURING THE DEPLOYMENT OF THIS POLICY

For the application of this and all other policies, general principles apply. (The general principles are given in 7.1–7.7 of P-01.)

7.2.   Important guidelines should be interpreted by a company position. Obviously, the company must define the term important in this context. For internationally operating companies, the definition will be different from that of companies operating in a single market. However it should be noted that guidelines applying worldwide, such as World Health Organization (WHO) guidelines, or to the major markets, such as International Conference on Harmonisation (ICH) guidelines, have considerable influence on the rule making in other countries; therefore, knowledge of these guidelines helps to adapt in-house standards in time and to foresee developments.

7.3.1. So far, there are several marketed products that offer segments of regulations in electronic format. It is suggested that each company carefully monitor the market in order to purchase this service. In the future, the Internet will possibly provide all U.S. and EU guidelines. At present, this does not seem to be a practical solution, as queries and downloading are a time-consuming process. With regard to international operations, there are no CROs that offer the complete guidelines worldwide.

7.3.2. Texts should be evaluated/interpreted by the respective scientific discipline(s) and Regulatory Affairs in order to capture both viewpoints. Also bear in mind that within the Regulatory Bodies, both scientists and regulators have worked on the document; so in addition to the scientific aspect, the procedural and political aspects must also be evaluated. The consequences to the company and its medicinal products should also be discussed.

7.3.3. This applies primarily to draft documents.

7.3.4. Though time- and capacity-consuming, the department/company should consider actively participating in working parties, as this provides a real opportunity to bring your ideas to the discussion process. Once a draft is released for comments, it may be difficult to change it significantly, unless, of course, there are major flaws in it.

A Policy of Department XYZ

26: Regulatory Strategy

Document Type:	Policy
Document Code:	26
	(enter company-specific code)
Title:	Regulatory Strategy
Date/Revision No.:	DD/MM/YY number xy
Scope:	Global
References:	(enter policies, standards, SOPs of your department/company, or other documents [e.g., guidelines] that should be considered in this context)
Authorization:	
	Signature of authorized person(s)
	Name of authorized person(s)
	Job title/Function of authorized person(s)
Issue Date:	DD/MM/YY
Implementation Date:	DD/MM/YY

1. PURPOSE

This policy is a set of rules developed to govern the generation and maintenance of the regulatory strategy by Regulatory Affairs worldwide.

2. DEFINITIONS

The key terms pertaining to this policy should be defined here. As there are no uniform and globally accepted definitions available, please develop your own definitions. In this way, the language of the staff of your organization can be incorporated. The term *regulatory strategy* signifies the selection of the appropriate submission strategy in terms of the content and presentation of dossier(s), as well as time point(s) for submission(s), and procedure(s) to be used, and considering the target summary of product characteristics (SMPC) for the developmental or marketed medicinal product and the regulatory environment.

3. STATEMENT OF POLICY

This section covers the actual rulings that should be complied with when working according to the principles of the quality system. For the topic *regulatory strategy,* rulings should be available for the following items:

3.1.   Responsibility for the generation and maintenance of the regulatory strategy is with the appointed Regulatory Affairs manager.

3.2.   The generation of the regulatory strategy starts with the official decision of the company to develop a medicinal product and is maintained until the company's target has been reached. It is regularly presented to the project team for discussion.

4. **RESPONSIBILITY FOR POLICY IMPLEMENTATION:**
   NAME AND TITLE/FUNCTION OF AUTHORIZED PERSON(S)

5. **RULING PERTINENT TO EXISTING PROCEDURES**

6. **RULING APPLICABLE IN THE CASE OF OUTSOURCING**

7. **POINTS TO CONSIDER**
   DURING THE DEPLOYMENT OF THIS POLICY

For the application of this and all other policies, general principles apply. (The general principles are given in 7.1–7.7 of P-01.)

7.3.1. One of the most interesting and challenging tasks for the Regulatory Affairs professional is to develop and maintain a regulatory strategy for a given medicinal product. It requires continuous and careful overview and evaluation of the actual regulatory environment as set out by regulations and guidelines in the light of experience. The regulatory strategy should also be discussed within Regulatory Affairs in order to use all available experience. Also consider contacting the Regulatory Bodies in advance to obtain their input on critical issues.

7.3.2. When discussing the regulatory strategy with the project team, it is important to first discuss all of the targets identified by Clinical and Marketing. Are all of the indications equally important? Is the exact wording essential? Are all of the markets to be accessed at the same time? It is important to develop jointly the most beneficial strategy for the company. Clearly give to all participants the benefits and risks for each scenario. Involve the project team members, as decisions on the regulatory strategy should be supported by the whole team.

Good documentation of agreements reached (and changes) is required, as product development and registration are lengthy processes and the team members may change. Documentation of the discussion process will also help to avoid going round in circles.

A Policy of Department XYZ

27: Submission

Document Type:	Policy
Document Code:	27
	(enter company-specific code)
Title:	Submission
Date/Revision No.:	DD/MM/YY number xy
Scope:	Global
References:	(enter policies, standards, SOPs of your department/company, or other documents [e.g., guidelines] that should be considered in this context)
	1. Policy on Dossier (P-11)
	2. Policy on Global Dossier (P-15)
	3. Policy on Documents for Regulatory Purposes (P-10)
	4. Submission Assignments Listing
	5. Departmental operational procedures
Authorization:	
	Signature of authorized person(s)
	Name of authorized person(s)
	Job title/Function of authorized person(s)
Issue Date:	DD/MM/YY
Implementation Date:	DD/MM/YY

1. PURPOSE

This policy is a set of rules developed to govern the generation of a submission by Regulatory Affairs worldwide.

2. DEFINITIONS

The key terms pertaining to this policy should be defined here. As there are no uniform and globally accepted definitions available, please develop your own definitions. In this way, the language of the staff of your organization can be incorporated.

- The term *dossier* signifies a compilation of documents for a specific regulatory purpose (e.g., application for clinical trial authorization or application for marketing authorization) in a specified country(ies) for a developmental or already marketed medicinal product in a structured form (i.e., submission-like). If applicable, it is a subset of the global dossier. The dossier is the basis for the submission(s).

- The term *global dossier* signifies a compilation of all documents required for international regulatory purpose(s) for a developmental or already marketed medicinal product. It is maintained continuously throughout the life cycle of the medicinal product and serves as a repository for the generation of dossiers and submissions.

- The term *submission* signifies a country-specific compilation of documents for a specific regulatory purpose (e.g., application for clinical trial authorization or application for marketing authorization) for a developmental or already marketed medicinal product in a structured form according to national regulatory requirements. It is based on the dossier, or, if applicable, the global dossier. It may contain additional national documents (e.g., national leaflets or application forms).

3. STATEMENT OF POLICY

This section covers the actual rulings that should be complied with when working according to the principles of the quality system. For the topic *submission,* rulings should be available for the following items:

3.1. The submission is based on the dossier (1), or, if applicable, the global dossier (2). The structure, format, and content of elements are defined by internal company standards (3).

3.2. Responsibility and single point of reference for the generation and documentation of the submission is with the assigned Regulatory Affairs manager. Assignments are documented in the submission assignments listing (4).

3.3. The submission is generated after an official decision is made to submit for a medicinal product in a specific country.

3.4. The submission must be generated in a timely fashion.

3.5. Each organizational unit responsible for generating a submission provides a written procedure for processes and responsibilities (5).

4. RESPONSIBILITY FOR POLICY IMPLEMENTATION: NAME AND TITLE/FUNCTION OF AUTHORIZED PERSON(S)

5. RULING PERTINENT TO EXISTING PROCEDURES

6. RULING APPLICABLE IN THE CASE OF OUTSOURCING

7. POINTS TO CONSIDER DURING THE DEPLOYMENT OF THIS POLICY

For the application of this and all other policies, general principles apply. (The general principles are given in 7.1–7.7 of P-01.)

7.1. One of the most important Regulatory Affairs functions is to generate dossiers for submission for regulatory purposes (e.g., to obtain a clinical trial license, a marketing authorization, or a renewal of a marketing authorization). The technical part of this process is reflected by the policies on global dossier, dossier, and submission. For companies operating in a single market, the policies on dossier and submission should be combined. (Small, unstructured submissions, e.g., single documents or statements, are not covered by this policy).

7.2. For companies continuously maintaining a global dossier, it may be identical with the dossier. However, most companies provide local Regulatory Affairs units with only the relevant subset for the intended submission, not the entire documentation. Usually, the dossier is provided in a submission-like structure to facilitate the generation of the submission. Based on your organization's needs, define whether the dossier will or will not be paginated and referenced. The submission may also contain locally required documents (e.g., application forms or national leaflet texts).

7.3.1. For companies operating in a single market, the dossier may be identical to the submission.

7.3.2. Develop the submission assignments listing.

7.3.4. Define the time frame for your organization. A suggested time frame is one month or less after the receipt of the dossier, or, if applicable, the global dossier.

7.3.5. Develop the SOP(s) for generating a submission.

A Policy of Department XYZ

28: Terminology

Document Type:	Policy
Document Code:	28
	(enter company-specific code)
Title:	Terminology
Date/Revision No.:	DD/MM/YY number xy
Scope:	Global
References:	(enter policies, standards, SOPs of your department/company, or other documents [e.g., guidelines] that should be considered in this context)
	1. Standard for Terminology (S-28.01)
Authorization:	
	Signature of authorized person(s)
	Name of authorized person(s)
	Job title/Function of authorized person(s)
Issue Date:	DD/MM/YY
Implementation Date:	DD/MM/YY

1. PURPOSE

This policy is a set of rules developed to specify the terminology to be used by the staff of Regulatory Affairs worldwide. The objective of the policy is to ensure effective communication in order to achieve and maintain the required high standard of quality in terms of the function of Regulatory Affairs within the company worldwide.

2. DEFINITIONS

The key terms pertaining to this policy should be defined here. As there are no uniform and globally accepted definitions available, please develop your own definitions. In this way, the language of the staff of your organization can be incorporated. The term *terminology* in this policy means expressions (including abbreviations, if applicable) frequently used within Regulatory Affairs and/or requiring definition to clarify and standardize the meaning.

3. STATEMENT OF POLICY

This section covers the actual rulings that should be complied with when working according to the principles of the quality system. For the topic *terminology,* rulings should be available for the following items:

3.1.   Terminology used by Regulatory Affairs should be clear and not misleading.

3.2.   Terminology will be fixed as needed by a standard supplying an index with the long form, abbreviation if applicable, and a definition if necessary (1).

4. RESPONSIBILITY FOR POLICY IMPLEMENTATION: NAME AND TITLE/FUNCTION OF AUTHORIZED PERSON(S)

5. RULING PERTINENT TO EXISTING PROCEDURES

6. RULING APPLICABLE IN THE CASE OF OUTSOURCING

7. POINTS TO CONSIDER DURING THE DEPLOYMENT OF THIS POLICY

For the application of this and all other policies, general principles apply. (The general principles are given in 7.1–7.7 of P-01.)

7.3.2. Check and elaborate on the suggested standard. Add any terms you require in everyday work.

A Policy of Department XYZ

29: Tools

Document Type:	Policy
Document Code:	29
	(enter company-specific code)
Title:	Tools
Date/Revision No.:	DD/MM/YY number xy
Scope:	Global
References:	(enter policies, standards, SOPs of your department/company, or other documents [e.g., guidelines] that should be considered in this context)
	1. Policy on Information Technology (P-18)
	2. Policy on Information Management (P-17)
	3. Policy on Electronic Submission (P-13)
	4. Policy on Education/Training (P-12)
Authorization:	
	Signature of authorized person(s)
	Name of authorized person(s)
	Job title/Function of authorized person(s)
Issue Date:	DD/MM/YY
Implementation Date:	DD/MM/YY

1.   PURPOSE

This policy is a set of rules developed to govern the choice and use of tools by Regulatory Affairs worldwide.

2.   DEFINITIONS

The key terms pertaining to this policy should be defined here. As there are no uniform and globally accepted definitions available, please develop your own definitions. In this way, the language of the staff of your organization can be incorporated. The term *tools* signifies all programs or databases designed/purchased for the purpose of facilitating specific functions of Regulatory Affairs.

3.   STATEMENT OF POLICY

This section covers the actual rulings that should be complied with when working according to the principles of the quality system. For the topic *tools,* rulings should be available for the following items:

3.1.   The market will continuously be monitored for solutions/improvements concerning tools.

3.2.   Tools will be established only as needed. Policies on information technology (1), information management (2), and electronic submission (3) apply.

3.3.   Only validated tools will be used by Regulatory Affairs.

3.4.   Employees will receive adequate training. The policy on education/training (4) applies.

4.   RESPONSIBILITY FOR POLICY IMPLEMENTATION: NAME AND TITLE/FUNCTION OF AUTHORIZED PERSON(S)

5.   RULING PERTINENT TO EXISTING PROCEDURES

6.   RULING APPLICABLE IN THE CASE OF OUTSOURCING

7.   POINTS TO CONSIDER DURING THE DEPLOYMENT OF THIS POLICY

For the application of this and all other policies, general principles apply. (The general principles are given in 7.1–7.7 of P-01.)

7.1.   Tools might be beneficial in the areas of marketing history/registration status, the tracking of documents/dossiers, document management, the compilation of dossiers, archiving/retrieval, product information, regulations, and guidelines.

7

Standards

This chapter provides the standards as definitions of items that must be the same throughout the organization. Standards are established to assure that policies are fully implemented. For background information on the topics of standards, see the relevant policy in chapter 6, "Policies"; for more extensive background information, see chapter 5, "The Philosophy Behind the Policies".

The reader is invited to adapt the policies and/or standards to his or her organization and function and, if required, to develop quality assurance processes. These processes should form the basis of Standard Operating Procedures (SOPs). If required, further standards and SOPs may be established (if required separately, also on a national level); however, they must not deviate from existing policies or standards.

A Department XYZ Standard

01.01: Policy

Document Type:	Standard
Document Code:	01.01
	(enter company-specific code)
Title:	Policy
Date/Revision No.:	DD/MM/YY number xy
Scope:	Global
References:	(enter policies, standards, SOPs of your department/company, or other documents [e.g., guidelines] that should be considered in this context)
	implements Policy on Policy (P-01)
Authorization:	
	Signature of authorized person(s)
	Name of authorized person(s)
	Job title/Function of authorized person(s)
Issue Date:	DD/MM/YY
Implementation Date:	DD/MM/YY

TEMPLATE AND CONTENT EXPLANATION
FOR POLICY COVER PAGE

A Policy of Department XYZ

XX: XXXXX *(document code and title)*

Document Type:	Policy
Document Code:	*(two-digit numbering)*
	(enter company-specific code)
Title:	*(document title—brief)*
Date/Revision No.:	(e.g., June 96 / rev 2)
Scope:	*(functional units and/or regional or geographical areas affected, e.g., Global)*
References:	(enter policies, standards, SOPs of your department/company, or other documents [e.g., guidelines] that should be considered in this context)
	(references to other documents; namely, other
	Policies, Standards, or SOPs, especially from other
	functional units; P = Policy, S = Standard, SOP =
	Standard Operating Procedure [e.g., SOP-04])
Authorization:	*(authorization is determined for each type of document by the concerned functional unit)*
	Signature of authorized person(s)
	Name of authorized person(s)
	Job title/Function of authorized person(s)
Issue Date:	*(date of [last] signature)*
Implementation Date:	*(date by which policy must be implemented)*

P-XX   page 1 of XX

TEMPLATE AND STANDARD TEXT
FOR BODY OF POLICY

---

1.   PURPOSE

This policy is a set of rules developed to . . .

2.   DEFINITIONS

The key terms pertaining to this policy should be defined here. As there are no uniform and globally accepted definitions available, please develop your own definitions. In this way, the language of the staff of your organization can be incorporated.

3.   STATEMENT OF POLICY

This section covers the actual rulings that should be complied with when working according to the principles of the quality system. For the topic (insert subject of policy), such ruling should be available for the following items:

4.   RESPONSIBILITY FOR POLICY IMPLEMENTATION:
NAME AND TITLE/FUNCTION OF AUTHORIZED PERSON(S)

5.   RULING PERTINENT TO EXISTING PROCEDURES

6.   RULING APPLICABLE IN THE CASE OF OUTSOURCING

7.   POINTS TO CONSIDER
DURING THE DEPLOYMENT OF THIS POLICY

For the application of this and all other policies, general principles apply. (The general principles are given in 7.1–7.7 of P-01.)

P-XX   page 2 of XX

A Department XYZ Standard

01.02: Standard Operating Procedure

Document Type:	Standard
Document Code:	01.02
	(enter company-specific code)
Title:	Standard Operating Procedure
Date/Revision No.:	DD/MM/YY number xy
Scope:	Global
References:	(enter policies, standards, SOPs of your department/company, or other documents [e.g., guidelines] that should be considered in this context)
Authorization:	
	Signature of authorized person(s)
	Name of authorized person(s)
	Job title/Function of authorized person(s)
Issue Date:	DD/MM/YY
Implementation Date:	DD/MM/YY

TEMPLATE AND CONTENT EXPLANATION
FOR SOP COVER PAGE

A Department XYZ Standard

XX: XXXXX *(document code and title)*

Document Type:	Standard
Document Code:	*(two-digit numbering of the policy plus the two-digit numbering of the standard, separated by a decimal, e.g., 01.02)*
	(enter company-specific code)
Title:	*(document title—brief)*
Date/Revision No.:	(e.g., June 96 / rev 2)
Scope:	*(functional units and/or regional or geographical areas affected, e.g., Global)*
References:	(enter policies, standards, SOPs of your department/company, or other documents [e.g., guidelines] that should be considered in this context)
	(references to other documents; namely, other
	Policies, Standards, or SOPs, especially from other
	functional units; P = Policy, S = Standard, SOP =
	Standard Operating Procedure [e.g., SOP-04])
Authorization:	*(authorization is determined for each type of document by the concerned functional unit)*
	Signature of authorized person(s)
	Name of authorized person(s)
	Job title/Function of authorized person(s)
Issue Date:	*(date of [last] signature)*
Implementation Date:	*(date by which policy must be implemented)*

SOP-XX   page 1 of XX

TEMPLATE AND STANDARD TEXT
FOR BODY OF SOP

1.   PURPOSE

This Standard Operating Procedure (SOP) covers the work associated with
. . .

2.   DEFINITIONS

The key terms pertaining to this SOP should be defined here. As there are no
uniform and globally accepted definitions available, please develop your own
definitions. In this way, the language of the staff of your organization can be incorporated.

3.   STATEMENT OF SOP

This section covers the actual procedures that should be complied with when
working according to the principles of the quality system. For the topic (insert
subject of SOP), procedures should be available for the following items:

4.   RESPONSIBILITY FOR SOP IMPLEMENTATION:
NAME AND TITLE/FUNCTION OF AUTHORIZED PERSON(S)

5.   RULING PERTINENT TO EXISTING PROCEDURES

6.   RULING APPLICABLE IN THE CASE OF OUTSOURCING

SOP-XX   page 2 of XX

A Department XYZ Standard

03.01: U.S. Application for Clinical Trial License: IND Content and Format

Document Type:	Standard
Document Code:	03.01
	(enter company-specific code)
Title:	U.S. Application for Clinical Trial License:
	IND Content and Format
Date/Revision No.:	DD/MM/YY number xy
Scope:	Global
References:	(enter policies, standards, SOPs of your department/company, or other documents [e.g., guidelines] that should be considered in this context)
	1. Code of Federal Regulations 21 § 312.23
	(1996 Edition)
Authorization:	
	Signature of authorized person(s)
	Name of authorized person(s)
	Job title/Function of authorized person(s)
Issue Date:	DD/MM/YY
Implementation Date:	DD/MM/YY

U.S. Application for Clinical Trial License: IND Content and Format

See also Code of Federal Regulations 21 § 312.23 (1996 Edition) (1)

1	Cover sheet (Form FDA 1571)
1.i	Name, address, and telephone number of sponsor; date of application; name of investigational new drug
1.ii	Identification of the phase or phases of the clinical investigation to be conducted
1.iii	Commitment not to begin clinical trials until an IND covering them is in effect
1.iv	Commitment that an Institutional Review Board (IRB) will be responsible for initial and continuing review and approval of each study
1.v	Commitment to conduct the investigation in accordance with all other applicable regulatory requirements
1.vi	Name and title of the monitor
1.vii	Name(s) and title(s) of person(s) responsible for review and evaluation of information relevant to the safety of the drug
1.viii	CRO's name and address in case of obligations transferred and a listing of these
1.ix	Signature(s)
2	Table of contents
3	Introductory statement and general investigational plan
3.i	Introductory statement giving name of drug and active ingredient(s), drug's pharmacological class, structural formula (if known), formulation of dosage form(s) to be used, route of administration, broad objectives, and planned duration of the proposed clinical investigation(s)
3.ii	Brief summary of previous human experience with the drug, with reference to other INDs if pertinent, and to investigational or marketing experience in other countries that may be relevant to the safety of the proposed clinical investigation(s)

3.iii	Countries where the drug was withdrawn from investigation or marketing and the reason for withdrawal
3.iv	Overall plan for investigating the drug product for the following year, including:
3.iv.a	Rationale for the drug or research study
3.iv.b	Indication(s) to be studied
3.iv.c	General approach to be followed in evaluation
3.iv.d	Kinds of clinical trials to be conducted in the first year following submission (Indicate if plans are not developed for the entire year)

| 4 | [Reserved] |

5	Investigator's brochure, containing
5.i	Brief description of drug substance, formulation, and structural formula (if known)
5.ii	Summary of the pharmacological and toxicological effects of the drug in animals and, to the extent known, in humans
5.iii	Summary of the pharmacokinetics and biological disposition of the drug in animals and, if known, in humans
5.iv	Summary of information relative to safety and effectiveness in humans
5.v	Description of possible risks and side effects to be anticipated on the basis of prior experience with the drug under investigation or with related drugs, and of precautions or special monitoring to be done as part of the investigational use of the drug

6	Protocols
6.i	for each planned study
6.ii	describing all aspects of the study
6.iii	A protocol is required to contain:
6.iii.a	Statement of the objectives and purpose of the study
6.iii.b	Name, address, and a statement of the qualifications (CV or other) of each investigator and subinvestigator
6.iii.c	Criteria for patient selection and exclusion, estimate number of patients

6.iii.d	Description of the design of the study including control to be used
6.iii.e	Method for determining the dose(s)
6.iii.f	Description of the observations and measurements in order to fulfill the objectives of the study
6.iii.g	Description of clinical procedures, laboratory tests, or other measures

7	Chemistry, manufacturing, and control information
7.i	Section describing the composition, manufacture, and control of the drug substance and the drug product
7.ii	Information to be submitted depends on the scope of the proposed clinical investigation
7.iii	Information amendments to supplement the initial information concerning chemistry, manufacturing, and control processes (if applicable)
7.iv	Reflecting the distinctions described in (a)(7), the submission is required to contain the following:
7.iv.a	Drug substance
	Description incl. physical, chemical, or biological characteristics; name and address of manufacturer; method of preparation, acceptable limits, and analytical methods to assure identity, strength, quality, purity; stability; if applicable reference(s) to USP-NF
7.iv.b	Drug product
	Components used in manufacture; quantitative composition incl. variations; name and address of manufacturer; manufacturing and packaging procedure; acceptable limits and analytical methods used to assure identity, strength, quality, purity; stability; if applicable reference(s) to USP-NF
7.iv.c	Brief general description of the composition, manufacture, and control of any placebo
7.iv.d	Labeling
7.iv.e	Environmental analysis requirements

8	Pharmacology and toxicology information
8.i	Pharmacology and drug disposition
8.ii	Toxicology
8.ii.a	Integrated summary of the toxicological effects of the drug in animals and in vitro
8.ii.b	Full tabulation of data suitable for detailed review (each toxicological study)
8.iii	GLP compliance statement (each nonclinical laboratory study subject to GLP regulations under 21 CFR Part 58)
9	Previous human experience with the investigational drug (summary)
9.i	Detailed information on previous experience
9.ii	Information acc. to (a)(9)(i) is required for each active drug component
9.iii	List of countries in which the drug has been marketed or withdrawn
10	Additional information
10.i	Drug dependence and abuse potential
10.ii	For radioactive drugs, sufficient data to calculate radiation-absorbed dose
10.iii	Other information
11	Relevant information on request
11.b	Reference to information previously submitted by filename, reference number, volume, and page number in agency records
11.c	Accurate and complete English translation of each part of the application that is not in English, plus for foreign language literature publications a copy of such publication
11.d	Number of copies: one original and two copies plus all amendments and reports
11.e	Numbering of IND submissions serially using a single, three-digit serial number; amendments, reports, and correspondence to be numbered chronologically in sequence

A Department XYZ Standard

04.01: EU Application for Marketing Authorization: Chemical Active Substance(s)

Document Type:	Standard
Document Code:	04.01
	(enter company-specific code)
Title:	EU Application for Marketing Authorization:
	Chemical Active Substance(s)
Date/Revision No.:	DD/MM/YY number xy
Scope:	Global
References:	(enter policies, standards, SOPs of your department/company, or other documents [e.g., guidelines] that should be considered in this context)
	1. Volume IIB, final (1/97)
Authorization:	
	Signature of authorized person(s)
	Name of authorized person(s)
	Job title/Function of authorized person(s)
Issue Date:	DD/MM/YY
Implementation Date:	DD/MM/YY

EU Application for Marketing Authorization:
Chemical Active Substance(s)

According to: Volume IIB: The Notice to Applicants—Presentation and
Content of the application dossier, final, January 1997.

I	Summary of the Dossier
I	Overall Table of Contents
I.A	Administrative Data/Marketing Authorization Particulars
	Fees/declaration and signature
	Proof of payment
	Type of application
	Marketing Authorization Particulars
	Manufacturers Authorization(s)
	Justification for use of one or more trade names in the Member States, if appropriate
	List of Samples sent with the Application
I.B	Summary of Product Characteristics (SMPC)
I.B.1	Summary of Product Characteristics
I.B.2	Proposals for Packaging, Labeling, and Package Leaflet
	Packaging
	Labeling
	Package Insert(s)
I.B.3	SMPCs already approved in the Member States
I.C	Expert Reports
I.C.1	Expert Report Chemical/Pharmaceutical/Biological
	1. Product Profile
	2. Expert Report
	3. Report Formats
	4. Written Summaries

I.C.2	Expert Report Pharmaco-Toxicological
	1. Product Profile
	2. Expert Report
	3. Tabulated Study Reports
	4. Written Summaries
I.C.3	Expert Report Clinical
	1. Product Profile
	2. Expert Report
	3. Tabulated Study Reports
	4. Written Summaries
II	Chemical, Pharmaceutical, and Biological Documentation
II	Table of Contents
II.A	Composition of the medicinal product
II.A.1	Formula
II.A.2	Container
II.A.3	Clinical trial formula(e)
II.A.4	Development Pharmaceutics
II.B	Method of preparation
II.B.1	Manufacturing formula
II.B.2	Manufacturing process
II.B.3	Validation of the Process
II.C	Control of starting materials
II.C.1	Active substance(s)
II.C.1.1	Specifications and Routine Tests
II.C.1.1.1	Active substance(s) described in a Pharmacopoeia
II.C.1.1.2	Active substance(s) not described in a Pharmacopoeia
II.C.1.1.2.-	Characteristics
II.C.1.1.2.-	Identification Tests
II.C.1.1.2.-	Purity Tests

A Department XYZ Standard

04.02: EU Application for Marketing Authorization: Biological(s), Part II

Document Type:	Standard
Document Code:	04.02
	(enter company-specific code)
Title:	EU Application for Marketing Authorization:
	Biological(s), Part II
Date/Revision No.:	DD/MM/YY number xy
Scope:	Global
References:	(enter policies, standards, SOPs of your department/company, or other documents [e.g., guidelines] that should be considered in this context)
	1. Volume IIB, final (1/97)
	2. Standard for EU Application for Marketing Authorization: Chemical Active Substance(s) (S-04.01)
Authorization:	
	Signature of authorized person(s)
	Name of authorized person(s)
	Job title/Function of authorized person(s)
Issue Date:	DD/MM/YY
Implementation Date:	DD/MM/YY

EU Application for Marketing Authorization: Biologicals, Part II (for other parts see S-04.01)

According to: Volume IIB: The Notice to Applicants—Presentation and Content of the application dossier, final, January 1997 (1). For other parts of dossier, see S-04.01 (2).

II.C.1.4.c	Protocol for preparation of all subsequent new WCBs, from the MCB
II.C.1.5	Production
II.C.1.5.1	Fermentation and harvesting
II.C.1.5.1.a	Name(s) and address(es) of the production site(s)
II.C.1.5.1.b	Definition of a batch
II.C.1.5.1.c	Flow diagram (annotated)
II.C.1.5.1.d	Brief description of equipment and facilities
II.C.1.5.1.e	Composition, preparation, sterilization, specification, and storage of culture media and other additives
II.C.1.5.1.f	Storage of the intermediate harvests
II.C.1.5.1.g	In-process controls including criteria for acceptance of each harvest
II.C.1.5.1.j	Pilot scale fermentation if different in procedure, or scale to that proposed to be marketed batches
II.C.1.5.2	Purification
II.C.1.5.2.a	Name(s) and address(es) of the production site(s)
II.C.1.5.2.b	Definition of a batch
II.C.1.5.2.c	Flow diagram (annotated)
II.C.1.5.2.d	Brief description of equipment and facilities
II.C.1.5.2.e	Composition, preparation, sterilization, and specifications of reagents, buffers, eluents, and other chemicals
II.C.1.5.2.f	Conditions of use and reuse of columns, including loading conditions, yields, regeneration and storage between runs, lifetime of each column
II.C.1.5.2.g	Storage of intermediates
II.C.1.5.2.h	In-process controls including elution profiles, limits, and criteria for selection and acceptance of the desired fraction for each chromatographic step
II.C.1.5.2.i	Reprocessing criteria for each purification step
II.C.1.5.2.j	Pilot scale purification if different in procedure or scale to that proposed to be marketed batches

A Department XYZ Standard

04.03: U.S. Application for Marketing Authorization

Document Type:	Standard
Document Code:	04.03
	(enter company-specific code)
Title:	U.S. Application for Marketing Authorization:
	NDA Content and Format
Date/Revision No.:	DD/MM/YY number xy
Scope:	Global
References:	(enter policies, standards, SOPs of your department/company, or other documents [e.g., guidelines] that should be considered in this context)
	1. Code of Federal Regulations 21 § 314.50
	(1996 edition)
Authorization:	
	Signature of authorized person(s)
	Name of authorized person(s)
	Job title/Function of authorized person(s)
Issue Date:	DD/MM/YY
Implementation Date:	DD/MM/YY

U.S. Application for Marketing Authorization: NDA Content and Format

See also Code of Federal Regulations 21 § 314.50 (1996 Edition) (1)

a	Application form
a.1	Name/address of applicant, date of application, application number if previously issued, name of drug product (and other names), dosage form, strength, route of administration, nos. of INDs referred to, nos. of DMFs referred to, proposed indications for use
a.2	Type of submission
a.3	Status (POM or OTC)
a.4	Overall table of contents
a.5	Signature(s)

b	Index

c.	Summary
c.1	Summary of application
c.2.i	Proposed labeling text with annotations
c.2.ii	Pharmacologic class/rational use/clinical benefits
c.2.iii	Marketing history
c.2.iv	Summary of chemistry/manufacturing/controls section
c.2.v	Summary of nonclinical pharmacology/toxicology section
c.2.vi	Summary of human pharmacokinetics and bioavailability section
c.2.vii	Summary of microbiology section (for anti-infective drugs)
c.2.viii	Summary of clinical data section
c.2.ix	Benefit/risk discussion and proposed postmarketing studies

d	Technical sections
d.1	Chemistry, manufacturing, and controls section
d.1.i	Drug substance: physical and chemical characteristics, stability, name and address of manufacturer, method of synthesis/

isolation, purification, in-process controls, specifications and analytical methods to assure identity, strength, quality, purity and bioavailability, incl. references to USP-NF if applicable

d.1.ii.a   Drug product: a list of all components used in the manufacture of the drug product, a statement of the composition of the drug product, a statement of the specifications and analytical methods for each component, name and address of each manufacturer, a description of the manufacturing and packaging procedures and in-process controls, incl. references to USP-NF if applicable

d.1.ii.b   Necessary information on the drug product (e.g., batch production record for stability and biobatches, specifications and test procedures for each component and the drug product)

d.1.ii.c   Proposed or actual master production record

d.1.iii   Environmental impact

d.1.iv   Possible submission of a complete chemistry, manufacturing, and controls section 90 to 120 days before remainder of application

d.1.v   Statement certifying that the field copy of the application has been provided to the applicant's home FDA district office

d.2   Nonclinical pharmacology/toxicology section

d.2.i   Pharmacology studies

d.2.ii   Toxicology studies

d.2.iii   Studies, as appropriate, of the effects of the drug on reproduction and on the developing fetus

d.2.iv   Any studies of the absorption, distribution, metabolism, and excretion of the drug in animals

d.2.v   For each nonclinical laboratory study subject to the GLP regulations, a statement that it was conducted in compliance with those regulations or a brief statement of the reason for the noncompliance

d.3   Human pharmacokinetics and bioavailability section

d.3.i   Description of each of the bioavailability and pharmacokinetic studies of the drug in humans performed by or on behalf of the applicant that includes a description of the analytical and statistical methods d.3.ii   If the application describes in the chemistry, manufacturing, and controls section specifications or analytical methods needed to assure the bioavailability of the drug product or substance or both, a statement in this section of the rationale for establishing the specification or analytical methods, including data and information supporting the rationale

d.3.iii   Summarizing discussion and analysis of the pharmacokinetics and metabolism of the active ingredient(s) and the bioavailability or bioequivalence, or both, of the drug product

d.4   Microbiology section

d.4.i   Description of the biochemical basis of the drug's action on microbial physiology

d.4.ii   Description of the antimicrobial spectra of the drug, including results of in vitro preclinical studies to demonstrate concentrations of the drug required for effective use

d.4.iii   Description of any known mechanisms of resistance to the drug including results of any known epidemiologic studies to demonstrate prevalence of resistance factors

d.4.iv   Description of clinical microbiology laboratory methods (for example, in vitro sensitivity discs) needed for effective use of drug

d.5   Clinical data section

d.5.i   Clinical pharmacology studies

d.5.ii   Controlled clinical studies

d.5.iii   Uncontrolled clinical studies

d.5.iv   Other studies and information

d.5.v   Integrated summary (effectiveness)

d.5.vi   Summary and updates of safety information

d.5.vi.a   Integrated summary (safety)

d.5.vi.b   Safety update

d.5.vii   Potential of abuse, drug abuse/overdose information

d.5.viii   Integrated summary of benefits/risks

d.5.ix   Compliance statements for studies in humans

d.5.x   Name and address of CROs, obligations transferred

d.5.xi	List identifying each clinical study where original subject records were audited or reviewed by the sponsor in the course of monitoring to verify the accuracy
d.6	Statistical section
d.6.i	Copy of the information submitted of each controlled clinical study [= d.5.ii]
d.6.ii	Copy of information submitted under clinical section, integrated summary safety [= d.5.vi.a]
e	Samples/labeling
e.1	Samples on request
e.1.i	4 representative samples in sufficient quantity for 3x every test described of the following:
e.1.i.a	Drug product
e.1.i.b	Drug substance used in the samples under e.1.i.a
e.1.i.c	Reference standards, blanks (except recognized pharmacopeial standards)
e.1.ii	Samples of the finished market package on request
e.2	Archival copy
e.2.i	3 copies of the analytical methods and related descriptive information d.1
e.2.ii	Copies of label and all labeling for drug product (4 copies of draft labeling or 12 copies of final printed labeling)
f	Case report forms and tabulations
f.1	Case report tabulations
f.2	Case report forms
f.3	Additional data
f.4	Meeting with FDA to discuss the presentation and format of supportive information (e.g., submission of tabulations of patient data and CRFs on microfiche or computer tapes)

g	Other
g.1	Reference to information previously submitted by filename, reference number, volume, and page number in the agency's records, for third party information LOA
g.2	Accurate and complete English translation of each part of the application that is not in English, plus for foreign language literature publications a copy of such publication
g.3	Right of reference to use an investigation (LOA)
h	Patent information
h.i	Patent certification
h.i.1	Contents
h.i.1.i	Patents claiming drug, drug product, or method of use
h.i.1.i.A	except as provided under i.2 for each U.S. patent, as applicable:
h.i.1.i.A.1	Paragraph I Certification—Patent information has not been submitted to FDA
h.i.1.i.A.2	Paragraph II Certification—patent has expired
h.i.1.i.A.3	Paragraph III Certification—expiry date for patent
h.i.1.i.A.4	Paragraph IV Certification—patent is invalid, unenforceable, or will not be infringed by the application
h.i.1.i.B	If reference drug is itself licensed generic of patented drug, as applicable
h.i.1.ii	No relevant patents
h.i.1.iii	Method of use patent
h.i.1.iii.A	Statement that the method of use patent does not claim as of the proposed indications
h.i.1.iii.B	If labeling includes an indication already covered by (use) patent
h.i.2	Method of manufacturing patent
h.i.3	Licensing agreements
h.i.4	Late submission of patent information
h.i.5	Disputed patent information
h.i.6	Amended certifications

h.i.6.i	After finding of infringement
h.i.6.ii	After removal of a patent from the list
h.i.6.iii	Other amendments
h.i.6.iii.A	Amendment to submitted certification if the certification is no longer accurate [except i.4 and i.6.iii.B]
h.i.6.iii.B	An applicant is not required to amend a submitted certification when information on an otherwise applicable patent is submitted after the effective date of approval
j	Claimed exclusivity
j.1	Statement of claimed exclusivity
j.2	Reference to the appropriate paragraph that supports the claim
j.3	Assurance that the drug has not previously been approved
j.4	Assurance that the application contains "new clinical investigations" that are "essential to approval of the application or supplement" and were "conducted or sponsored by the applicant"
j.4.i	New clinical investigations
j.4.ii	Essential to approval
j.4.iii	Conducted or sponsored by
k	Format of an original application
k.1	Complete archival copy
k.2	Review copy of the application
k.3	Field copy of the application containing the technical section, copy of application, copy of summary, and a certification that the field copy is a true one
k.4	Sufficient folders to bind the copies of the application from FDA

A Department XYZ Standard

07.01: Regulatory Body Contact Report

Document Type:	Standard
Document Code:	07.01
	(enter company-specific code)
Title:	Regulatory Body Contact Report
Date/Revision No.:	DD/MM/YY number xy
Scope:	Global
References:	(enter policies, standards, SOPs of your department/company, or other documents [e.g., guidelines] that should be considered in this context)
	implements Policy on Contact Report (P-07)
Authorization:	
	Signature of authorized person(s)
	Name of authorized person(s)
	Job title/Function of authorized person(s)
Issue Date:	DD/MM/YY
Implementation Date:	DD/MM/YY

Contact with:

    Regulatory Body contacted: _____

    Name, first name(s) of Person(s) contacted/present: _____

    Function/job title of person(s) contacted/present: _____

Date of Contact: DD/MM/YY
_____

Subject:

    Medicinal Product: (internal code, generic/trade, name if required, dose,
_____

                    formulation, indication)
_____

    Reference No.:   (e.g., IND, NDA, ANDA, internal code(s))
_____

    Keyword(s):   (e.g., stability)
_____

Date of Report: DD/MM/YY
_____

Brief summary of results of contact or actions to be taken

_____

_____

_____

Author(s) of Report:

    Signature(s) _____

    Name(s) _____

    Function/job title(s) _____

Distribution to: _____

A Department XYZ Standard

10.01: Regulatory Document Types

Document Type:	Standard
Document Code:	10.01
	(enter company-specific code)
Title:	Regulatory Document Types
Date/Revision No.:	DD/MM/YY number xy
Scope:	Global
References:	(enter policies, standards, SOPs of your department/company, or other documents [e.g., guidelines] that should be considered in this context)
	implements Policy on Documents for Regulatory
	Purposes (P-10)
Authorization:	
	Signature of authorized person(s)
	Name of authorized person(s)
	Job title/Function of authorized person(s)
Issue Date:	DD/MM/YY
Implementation Date:	DD/MM/YY

Annotated Package Insert

Annual Report (investigational medicinal product)

Application for Clinical Trial Authorization

Application for Manufacturing License

Application for Marketing Authorization

Application for Renewal of Marketing Authorization

Assessment Report

Bibliography

Clinical Trial Authorization

Contact Report

Contract

Correspondence with Regulatory Body (not covered by other document types)

Deficiency Letter/Questions from Regulatory Body

Dossier

Drug Master File

Export License

External correspondence (not covered by other document types)

Free Sales Certificate

Global Dossier (if applicable)

Import License

Information Amendment (investigational medicinal product)

Inspection Report

Internal correspondence

Labeling

Letter of Authorization

Manufacturing License

Marketing Authorization

Marketing History/Registration Status

Patient Information

Periodic Safety Update Report (marketed medicinal product)

Plant Master File

Product Profile

Professional Information

Promotion/advertising material

Protocol Amendment (investigational medicinal product)

Summary Basis of Approval

Supplement (marketed medicinal product)

Table of Contents

A Department XYZ Standard

11.01: Dossier

Document Type:	Standard
Document Code:	11.01
	(enter company-specific code)
Title:	Dossier
Date/Revision No.:	DD/MM/YY number xy
Scope:	Global
References:	(enter policies, standards, SOPs of your department/company, or other documents [e.g., guidelines] that should be considered in this context)
	1. Policy on Dossier (P-11)
	2. Policy on Documents for Regulatory Purposes (P-10)
	3. Departmental operational procedures
Authorization:	
	Signature of authorized person(s)
	Name of authorized person(s)
	Job title/Function of authorized person(s)
Issue Date:	DD/MM/YY
Implementation Date:	DD/MM/YY

1. PURPOSE

In order to meet the rules as set out in the policy on dossiers (1), and the policy on documents for regulatory purposes (2), quality is defined by this standard through specifiying quality of input and output and identifying key processes to be validated.

2. DEFINITIONS

The key terms pertaining to this policy should be defined here. As there are no uniform and globally accepted definitions available, please develop your own definitions. In this way, the language of the staff of your organization can be incorporated.

- The term *dossier* signifies a compilation of documents for a specific regulatory purpose (e.g., application for clinical trial authorization or application for marketing authorization) in a specified country(ies) for a developmental or already marketed medicinal product in a structured form (i.e., submission-like). If applicable, it is a subset of the global dossier. The dossier is the basis for the submission(s).

- The term *global dossier* signifies a compilation of all documents required for international regulatory purpose(s) for a developmental or already marketed medicinal product. It is maintained continuously throughout the life cycle of the medicinal product and serves as a repository for the generation of dossiers and submissions.

- The term *submission* signifies a country-specific compilation of documents for a specific regulatory purpose (e.g., application for clinical trial authorization or application for marketing authorization) for a developmental or already marketed medicinal product in a structured form according to national regulatory requirements. It is based on the dossier, or, if applicable, the global dossier. It may contain additional national documents (e.g., national leaflets or application forms).

3. STATEMENT OF STANDARD

1. Quality of Processes

The following processes (if applicable) are adequately standardized and validated (3): Selection/incorporation of documents for registration purposes into dossier, production of Table of Contents (TOC), copying from originals, printing originals from electronic archiving system(s), putting copies in order of TOC to produce master copy, paginating master copy, copying from master copy to produce copies for national Regulatory Affairs (RA), putting in binders, labeling, export to magneto-optical or compact disc, mailing dossier.

2. Quality of Input

Specs/Limits	Frequency/Extent of Checks	Documentation of Check Results	Responsibility for Checks	Duplication of Checks by RA
1. Original Documents				
1.1 Complete (all pages)	every doc: every p	monthly report to head RA, disciplines concerned	supplier (=archive) may be delegated to original supplier if adequately controlled	spot check on copies: all expert reports/summaries 1st 5 CMC documents 1st report - Pharmacology - Toxicology - Clinical 1st 2 other documents
1.2 Format and content acc. to state of the art, SOPs, GMP, GLP, GCP; acc. to internal company standards (2)	all items			
1.3 Good copying quality	every p			
1.4 Cross-references OK	all cross-references			
1.5 Good English	main part			
2. Copies from originals for dossier master copy				
2.1 Complete (all pages)	CMC every page (incl. attachments) other: spot check	monthly report to head RA	Copying department	spot check: all expert reports/summaries, all CMC docs, 3 reports, 2 other
2.2 Good copying quality	first p + spot check			

Table continued on next page.

Table continued from previous page.

Specs/Limits	Frequency/ Extent of Checks	Documentation of Check Results	Responsibility for Checks	Duplication of Checks by RA
3. Material				
3.1 Paper: Good copying quality	each delivery/1 p	yearly report to head RA	Copying department	None
3.2 Binders: stable, company Logo	each delivery/1 binder	yearly report to head RA	person responsible for ordering material	check on master copy
3.3 Labels: good adhesion	each delivery	yearly report to head RA	person responsible for ordering material	check on master copy
4. Contents of dossier	every dossier	monthly report to head RA		Head RA: spot checks
4.1 Structure, format and content acc. to internal company standards (2)	TOC/monthly			
4.3 Updates processed within 10 working days of availability of new documents to RA				
4.4 Changes and amendments adequately documented				
4.5 Acc. to SOP (3)				

Table continued on next page.

S-11.01   page 5 of 6

3. Quality of Output (= Dossier)

Specs/Limits	Frequency/ Extent of Checks	Documentation of Check Results	Responsibility for Checks	Duplication of Checks by RA
1. Dossier master copy				
1.1 Valid versions, most actual	every doc	final report to head RA	RA	National RA: spot check
1.2 Correct sequence acc. to standard	all docs all expert reports/ summaries			
1.3 For expert reports, summaries: Xrefs to documentation OK	all Xrefs			
1.4 CMC section harmonized	all CMC docs			
2. Copies (from dossier master copy) for countries				
2.1 Complete (all pages)	spot check (1 copy or 10% of copies, whichever is greatest, summary part all pages, 1st vol of other parts all pages)	yearly report to head RA	Copying department	RA: spot check: (1 copy or 10% of copies, whichever is greatest, summary part all pages, 1 vol of another part all pages)
2.2 Good copying quality	first p + spot check			

A Department XYZ Standard

15.01: Global Dossier

Document Type:	Standard
Document Code:	15.01
	(enter company-specific code)
Title:	Global Dossier
Date/Revision No.:	DD/MM/YY number xy
Scope:	Global
References:	(enter policies, standards, SOPs of your department/company, or other documents [e.g., guidelines] that should be considered in this context)
	1. Policy on Global Dossier (P-15)
	2. Policy on Documents for Regulatory Purposes
	(P-10)
	3. Departmental operational procedures
	4. Standard for Dossier (S-11.01)
Authorization:	
	Signature of authorized person(s)
	Name of authorized person(s)
	Job title/Function of authorized person(s)
Issue Date:	DD/MM/YY
Implementation Date:	DD/MM/YY

1. PURPOSE

In order to meet the rules as set out in the policy on global dossiers (1), and the policy on documents for regulatory purposes (2), quality is defined by this standard through specifiying quality of "raw material(s)" and "finished product(s)" and identifying key processes to be validated.

2. DEFINITIONS

The key terms pertaining to this policy should be defined here. As there are no uniform and globally accepted definitions available, please develop your own definitions. In this way, the language of the staff of your organization can be incorporated.

- The term *dossier* signifies a compilation of documents for a specific regulatory purpose (e.g., application for clinical trial authorization or application for marketing authorization), in a specified country(ies) for a developmental or already marketed medicinal product in a structured form (i.e., submission-like). If applicable, it is a subset of the global dossier. The dossier is the basis for the submission(s).

- The term *global dossier* signifies a compilation of all documents required for international regulatory purpose(s) for a developmental or already marketed medicinal product. It is maintained continuously throughout the life cycle of the medicinal product and serves as a repository for the generation of dossiers and submissions.

- The term *submission* signifies a country-specific compilation of documents for a specific regulatory purpose (e.g., application for clinical trial authorization or application for marketing authorization) for a developmental or already marketed medicinal product in a structured form according to national regulatory requirements. It is based on the dossier, or, if applicable, the global dossier. It may contain additional national documents (e.g., national leaflets or application forms).

3. STATEMENT OF STANDARD

1. Quality of Processes

The following processes (if applicable) are adequately standardized and validated (3): Selection/incorporation of documents into global dossier, production of Table of Contents (TOC), copying from originals, printing originals from electronic archiving system(s), putting copies in order of TOC to produce master copy, paginating master copy, copying from master copy to produce copies for regional/national Regulatory Affairs (RA), putting in binders, labeling, export to magneto-optical or compact disc, mailing global dossier.

2. Quality of Input

See standard for dossier (4).

3. Quality of Output

See standard for dossier (4).

Standard for Global Dossier

a   Index/General administrative information/Expert reports

b   Synopsis of application

c   Chemical, pharmaceutical, and biological documentation

d   Samples/labeling

e   Nonclinical/Toxicology, Pharmacology

f   Human Pharmacokinetics and Bioavailability Data

g   Microbiology Data

h   Clinical Data

j   Safety and other Update Reports

k   Statistical Data/Overall Summary

l   Cover sheet(s)

m   Raw Data

Explanatory Note: The global dossier has a tree-like structure. Levels are as indicated by the code. If available, the titles of the relevant section in the EU dossier and/or the U.S. dossier are indicated. "—" means that this title is not applicable in the respective dossier. Applicability of sections/requirements must be checked on a case-by-case basis for each medicinal product. Special requirements for biological, herbal, or radiopharmaceutical medicinal products have not been included.

Code*	Title	EU	U.S.
a	*Index/General Administrative Information/Expert Reports*	—	—
a.1	Introduction	—	—
a.1.1	Name/Address of Applicant, Date of Application, Name of Drug Product (and other Names), Dosage Form, Strength, Route, Nos. of INDs referred to, Nos. of DMFs referred to, Proposed Indications for Use	—	a.1 Name/Address of Applicant, Date of Application, Name of Drug Product (and other Names), Dosage Form, Strength, Route, Nos. of INDs referred to, Nos. of DMFs referred to, Proposed Indications for Use
a.1.2	Type of Submission	—	a.2 Type of Submission
a.1.3	Status (POM/OTC)	—	a.3 Status (POM/OTC)
a.1.4	Signatures	—	a.5 Signature(s)
a.1.5	Administrative Data	I.A Administrative Data	—
a.1.6	Type of application	I.A Type of application	
a.1.7	Proof of payment	I.A Proof of payment	
a.2	Overall Table of Contents	—	—
a.2.1	Overall Table of Contents NDA	—	a.4 Overall Table of Contents
a.2.2	Overall Table of Contents EU	I Overall Table of Contents	—
a.2.3	Overall Table of Contents EU (Biological)	—	—
a.3	Registration Status	—	—
a.3.1	Marketing History NDA	—	c.2.iii Marketing History

*Key: "a" = Index/General Administrative Information/Expert Reports

Table continued from previous page.

Code*	Title	EU	U.S.
a.3.2	Marketing Authorization(s) EU	Marketing Authorization(s) EU	—
a.3.2.1	Marketing Authorization from Member State of Origin and Summary of Product Characteristics approved by it	I.B.3 SMPCs already approved in the Member States	—
a.3.2.2	Copies of Marketing Authorizations granted by other Member States	Copies of Marketing Authorizations granted by other Member States	—
a.3.2.3	Third World Countries in which a Marketing Authorization is granted	Third World Countries in which a Marketing Authorization is granted	—
a.4	Expert Opinion Reports	—	—
a.4.2	Expert Reports EU (only evaluation)	I.C Expert Reports	—
a.4.2.0.1	Product Profile	1 Product Profile (appears in each Expert Report)	—
a.4.2.1	Expert Report on the Chemical, Pharmaceutical and Biological Documentation	I.C.1 Expert Report Chemical/ Pharmaceutical/Biological	—
a.4.2.1.1	Evaluation (Chemical, Pharmaceutical and Biological Documentation) (evaluation)	2 Expert Report	—
a.4.2.2	Expert Report on the Pharmaco-Toxicological Documentation	I.C.2 Expert Report Pharmaco-Toxicological	—
a.4.2.2.1	Evaluation (Pharmaco-Toxicological Documentation)	2 Expert Report	—

*Key: "a" = Index/General Administrative Information/Expert Reports

Table continued on next page.

S-15.01   page 6 of 45

Table continued from previous page.

Code*	Title	EU	U.S.
a.4.2.2.1.1	Evaluation part PD		
a.4.2.2.1.2	Evaluation part PK		
a.4.2.2.1.3	Evaluation part Tox		
a.4.2.3	Expert Report on the Clinical Documentation	I.C.3 Expert Report Clinical	—
a.4.2.3.1	Evaluation (Clinical Documentation)	2 Expert Report	—
a.5	Overall List of References	—	—
a.6	Overall Index	—	—
a.6.1	Index NDA	—	b Index
a.7	Other	—	—
b	***Synopsis of Application***		
b.1	Summary		—
b.1.1	Summary of Chemistry/Manufacturing/Controls Section		c.2.iv Summary of Chemistry/Manufacturing/Controls Section
b.1.1.1	Written Summary (Clinical, Pharmaceutical, and Biological Documentation)	4 Written Summaries	
b.1.1.2	Tabulated Summary (Clinical, Pharmaceutical, and Biological Documentation)	3 Report Formats	—
b.1.2	Summary of Nonclinical Pharmacology/Toxicology Section		—

Key: "a" = Index/General Administrative Information/Expert Reports; "b" = Synopsis of Application

Table continued from previous page.

Code*	Title	EU	U.S.
b.1.2.1	Written Summary (Pharmaco-Toxicological Documentation)	4 Written Summaries	c.2.v Summary of Nonclinical Pharmacology/Toxicology Section
b.1.2.1.1 1 of 2	Written Summary PD (Part)	—	Summary of Nonclinical Pharmacology and Toxicology Section (Part PD)
b.1.2.1.1 2 of 2	Written Summary PD (Part)	—	Nonclinical pharmacology summary
b.1.2.1.2 1 of 2	Written Summary PK (Part)	—	Summary of Nonclinical Pharmacology and Toxicology Section (Part PK)
b.1.2.1.2 2 of 2	Written Summary PK (Part)	—	Animal pharmacokinetic summary
b.1.2.1.3 1 of 2	Written Summary Tox (Part)	—	Summary of Nonclinical Pharmacology and Toxicology Section (Part Tox)
b.1.2.1.3 2 of 2	Written Summary Tox (Part)	—	Toxicology Summary
b.1.2.2	Tabulated Summary (Pharmaco-Toxicological Documentation)	3 Tabulated Study Reports	—
b.1.2.2.1	TSR PD	—	—
b.1.2.2.2	TSR PK	—	—
b.1.2.2.3	TSR Tox	—	—
b.1.3	Summary of Human pK, and Bioavailability Section/Clinical Data Section	4 Written Summaries	

Key: "b" = Synopsis of Application

Table continued on next page.

S-15.01   page 8 of 45

Table continued from previous page.

Code*	Title	EU	U.S.
b.1.3.1	Summary of Human pK and Bioavailability Section	—	c.2.vi Summary of Human Pharmaco-kinetic and Bioavailability Section
b.1.3.2	Summary of Clinical Data Section	—	c.2.vii Summary of Clinical Data Section
b.1.3.3	Tabulated Summary (Clinical)	3 Tabulated Study Reports	—
b.1.4	Summary of Microbiology Section		c.2.vii Summary of Microbiology Section
b.1.5	Benefit/Risk and proposed Postmarketing Studies	—	c.2.ix Benefit/Risk Discussion and proposed Postmarketing Studies
c	*Chemical, Pharmaceutical, and Biological Documentation*	—	—
c.0.1	General methods	—	—
c.0.1.1	General methods of determination	—	—
c.0.1.2	Assays	—	—
c.1	Table of Contents	—	—
c.1.2	Table of Contents EU	—	—
c.1.2.1	Table of Contents EU	II Table of Contents	—
c.2	Summary	—	—
c.3	Starting Materials	—	—
c.3.1	Active Ingredients	II.C.1 Active substance(s)	—

Key: "b" = Synopsis of Application; "c" = Chemical, Pharmaceutical, and Biological Documentation

Table continued on next page.

Table continued from previous page.

Code*	Title	EU	U.S.
c.3.1.1	Specifications and Routine Tests	II.C.1.1 Specifications and Routine Tests	Specifications and analytical methods for the drug substance Specifications (Bulk Drug) Active Substance Bulk Drug Monograph
c.3.1.1.1	Active Ingredients described in a Pharmacopoeia	II.C.1.1.1 Active Substance(s) described in a Pharmacopoeia	—
c.3.1.1.2	Active Ingredients not described in a Pharmacopoeia	II.C.1.1.2 Active Substance(s) not described in a Pharmacopoeia	—
c.3.1.1.2.1 1 of 2	Characteristics	II.C.1.2.- Characteristics	Summary of properties Physical and chemical properties
c.3.1.1.2.2	Identification Tests	II.C.1.2.- Identification Tests	Identification Tests
c.3.1.1.2.3	Purity Tests	II.C.1.2.- Purity Tests	Impurities and Degradation Product Determination of Heavy Metals
c.3.1.1.2.3.1	Physical	II.C.1.2.-. Physical	
c.3.1.1.2.3.2	Chemical	II.C.1.2.-. Chemical	
c.3.1.1.2.3.3	Biological/Immunological		
c.3.1.1.2.4	Other Tests	II.C.1.2.- Other Tests	Determination of Moisture by Karl Fischer reagent
c.3.1.2	Scientific data	II.C.1.2 Scientific data	
c.3.1.2.1	Nomenclature	II.C.1.2.1 Nomenclature	Structure and Nomenclature

Key: "c" = Chemical, Pharmaceutical, and Biological Documentation

Table continued on next page.

S-15.01   page 10 of 45

Table continued from previous page.

Code*	Title	EU	U.S.
c.3.1.2.1.1	Approved name(s)	—	
c.3.1.2.1.1.1	International Nonproprietary Name (INN)	II.C.1.2.1.- International Nonproprietary Name (INN)	
c.3.1.2.1.1.2	National Approved Name(s)	—	
c.3.1.2.1.2	Chemical name	II.C.1.2.1.- Chemical name	
c.3.1.2.1.3	Pharmacopeial Name(s)	—	
c.3.1.2.1.3.1	Ph. Eur. Name	—	
c.3.1.2.1.3.2	National Pharmacopeial Name	—	
c.3.1.2.1.4	Laboratory code	II.C.1.2.1.- Laboratory code	
c.3.1.2.1.5	Other name	II.C.1.2.1.- Other name	
c.3.1.2.2	Description	II.C.1.2.2 Description	Structure and Nomenclature
c.3.1.2.2.1	Physical form	II.C.1.2.2.- Physical form	
c.3.1.2.2.2	Structural formula (including conformational data for macromolecules)	II.C.1.2.2.- Structural formula (including conformational data for macromolecules)	
c.3.1.2.2.3	Molecular formula	II.C.1.2.2.- Molecular formula	
c.3.1.2.2.4	Relative molecular mass	II.C.1.2.2.- Relative molecular mass	
c.3.1.2.25	Chirality	II.C.1.2.2.- Chirality	

Key: "c" = Chemical, Pharmaceutical, and Biological Documentation

Table continued from previous page.

Code*	Title	EU	U.S.
c.3.1.2.3	Manufacture (Active Ingredient)	II.C.1.2.3. Manufacture	Method (s) of manufacture and packaging
c.3.1.2.3.1	Name(s) and address(es) of the manufacturing source(s)	II,C,1.2.3.- Name(s) and and address(es) of the manufacturing source(s)	
c.3.1.2.3.2	Synthetic or manufacturing route	II.C.1.2.3.- Synthetic or manufacturing route	Flow sheet
c.3.1.2.3.3	Description of process	II.C.1.2.3.- Description of process	Manufacturing procedure
c.3.1.2.3.4	Catalysts	II.C.1.2.3.- Catalysts	(Raw material Specifications)
c.3.1.2.3.5	Purification stages	II.C.1.2.3.- Purification Stages	
c.3.1.2.4	Quality control during manufacture	II.C.1.2.4. Quality control during manufacture	(In-process controls)
c.3.1.2.4.1	Starting materials	II.C.1.2.4.- Starting materials	(Raw Material Specifications)
c.3.1.2.4.2	Control tests on intermediate products	II.C.1.2.4.- Control tests on intermediate products	(In-process Controls)
c.3.1.2.5	Development	—	
c.3.1.2.5.1	Development chemistry	II.C.1.2.5 Development chemistry	
c.3.1.2.5.1.1	Evidence of chemical structure	II.C.1.2.5.- Evidence of chemical structure	
c.3.1.2.5.1.2	Potential isomerism	II.C.1.2.5.- Potential isomerism	(Polymorphism study)

Key: "c" = Chemical, Pharmaceutical, and Biological Documentation

Table continued on next page.

Table continued from previous page.

Code*	Title	EU	U.S.
c.3.1.2.5.1.3	Physicochemical characterization	II.C.1.2.5.- Physicochemical characterization	(Polymorphism study)
c.3.1.2.5.1.4	Full characterization of the primary reference material	II.C.1.2.5.- Full characterization of the primary reference material	(Primary Reference Standard)
c.3.1.2.6	Impurities	II.C.1.2.6 Impurities	
c.3.1.2.6.1	Potential impurities	—	
c.3.1.2.6.1.1	Potential impurities originating from the route of synthesis	II.C.1.2.6.- Potential impurities originating from the route of synthesis	
c.3.1.2.6.1.2	Potential impurities arising during the production and purification	II.C.1.2.6.- Potential impurities arising during the production and purification	
c.3.1.2.6.2	Analytical test procedures and their limits of detection	II.C.1.2.6.- Analytical test procedures and their limits of detection	
c.3.1.2.7	Batch analysis	II.C.1.2.7. Batch analysis	(Batch analysis)
c.3.1.2.7.1	Batches tested	II.C.1.2.7.- Batches tested	
c.3.1.2.7.2	Results of tests, including detailed data on the consistency of batches	—	
c.3.1.2.7.3	Reference material (analytical results), primary and others	II.C.1.2.7.- Reference material (analytical results), primary and others	(Primary Reference Standard)

Key: "c" = Chemical, Pharmaceutical, and Biological Documentation

Table continued on next page.

S-15.01   page 13 of 45

Table continued from previous page.

Code*	Title	EU	U.S.
c.3.1.2.7.3	Actual results of tests	II.C.1.2.7.- Results of tests	
c.3.2	Other Ingredients (incl. Ingredients no longer contained in the Finished Product)	II.C.2 Other Excipient(s)	Specifications and analytical methods for inactive components
c.3.2.1	Specifications and Routine Tests	II.C.2.1 Specifications and Routine Tests	
c.3.2.1.1	Ingredients described in a Pharmacopoeia	II.C.2.1.1 Excipient(s) described in a Pharmacopoeia	
c.3.2.1.2	Ingredients not described in a Pharmacopoeia	II.C.2.1.2.- Excipient(s) not described in a Pharmacopoeia	
c.3.2.1.2.1	Characteristics	II.C.2.1.2.- Characteristics	
c.3.2.1.2.2	Identification Tests	II.C.2.1.2.- Identification Tests	
c.3.2.1.2.3	Purity Tests	II.C.2.1.2.- Purity Tests	
c.3.2.1.2.3.1	Physical	II.C.2.1.2.- Physical	
c.3.2.1.2.3.2	Chemical	II.C.2.1.2.- Chemical	
c.3.2.1.2.4	Microbiological	—	
c.3.2.1.2.5	Other Tests	II.C.2.1.2.- Other Tests	
c.3.2.1.5.5	Assay(s) and /or evaluations	II.C.2.1.2.- Assay(s) and/or evaluations	
c.3.2.2	Scientific data	II.C.2.2 Scientific data	

Key: "c" = Chemical, Pharmaceutical, and Biological Documentation

Table continued on next page.

Table continued from previous page.

Code*	Title	EU	U.S.
c.3.2.2.1	Data, where necessary, for example on excipient(s) used for the first time in medicinal products	II.C.2.2 Data, where necessary, for example on excipient(s) used for the first time in medicinal products	
c.3.3	Container/Closure Information	—	
c.3.3.0.1	General Method of Determination for primary packaging	—	—
c.3.3.1	Summary	—	
c.3.3.2	Specifications and Routine Tests	II.C.3.1 Specifications and Routine Tests	
c.3.3.2.1	Container/Closure Specifications	—	Container/Closure Specifications
c.3.3.2.1.1 1 of 2	Type of Material	II.A.2 Container	
c.3.3.2.1.1 2 of 2	Type of Material	II.C.3.1.- Container (brief description)	
c.3.3.2.1.2	Construction	II.C.3.1.- Construction	
c.3.3.2.2	DMF Reference Letters	—	DMF Reference Letters
c.3.3.2.3	Monographs for Packaging Components/ Quality Specifications (routine tests and test procedures)	II.C.3.1.- Quality Specifications (routine tests and test procedures)	Monographs for Packaging Components
c.3.3.3	Scientific data	II.C.3.2. Scientific data	

Key: "c" = Chemical, Pharmaceutical, and Biological Documentation

Table continued on next page.

S-15.01   page 15 of 45

Table continued from previous page.

Code*	Title	EU	U.S.
c.3.3.3.1	Development Studies on Packaging Materials	II.C.3.2.- Development Studies on Packaging Materials	
c.3.3.3.2	Batch analysis, analytical results	II.C.3.2.- Batch analysis, analytical results	
c.3.4	Outer Packaging	—	—
c.3.4.1	Description	—	—
c.4	Intermediate Products		—
c.4.1	Manufacture/Controls	II.D Control Tests on Intermediate Products	
c.4.2	Stability		
c.5	Finished Product	—	—
c.5.1.1	Components of finished product		Components
c.5.1.2	Composition of finished product	II.A.1 Formula	Composition
c.5.1.2.1	Purity and Identification of Bulk Drug Substance Batches Used in Preclinical Safety and Clinical Studies	—	Purity and Identification of Bulk Drug Substance Batches Used in Preclinical Safety and Clinical Studies
c.5.1.2.2	Quantitative Composition of Dosage Form Used in Preclinical and Clinical Studies	—	Quantitative Composition of Dosage Form Used in Preclinical Safety and Clinical Studies
c.5.1.2.2.1	Dosage Forms used in Preclinical Studies	—	

*Key: "c" = Chemical, Pharmaceutical, and Biological Documentation

Table continued on next page.

S-15.01   page 16 of 45

Table continued from previous page.

Code*	Title	EU	U.S.
c.5.1.2.2.2	Dosage Forms used in Clinical Studies	II.A.3 Clinical trial formula(e)	Dosage Forms used in Clinical Studies
c.5.1.2.2.3	Consistency of the impurity profile of batches intended for marketing compared with that seen in all batches used in Preclinical and clinical studies (also possible as "c.3.1.2.6.9" under heading impurities, however can be written only at time of submission, therefore included here)	—	
c.5.1.2.3	Development Pharmaceutics		
c.5.1.2.3.1	Pharmaceutical Development	II.A.4 Development Pharmaceutics	(Preformulation Studies)
c.5.1.2.3.2	Analytical tests used for the pharmaceutical development of the product	II.Q Analytical tests used for the pharmaceutical development of the product	
c.5.2	Manufacture of Drug Product	—	
c.5.2.1	Manufacturers	—	Manufacturer(s)
c.5.2.2	Method(s) of manufacture and packaging	II.B.2 Manufacturing process	Method(s) of manufacture and packaging procedure and in-process controls
c.5.2.2.2	Process Flowchart		(Process Flowchart)
c.5.2.2.3	Standard Production Formula and Standard Operating Instructions	II.B.1 Manufacturing formula	Standard Production Formula and Standard Operating Instructions
c.5.2.2.4	Batch Record	—	(Batch Record)

Key: "c" = Chemical, Pharmaceutical, and Biological Documentation

Table continued on next page.

S-15.01   page 17 of 45

Table continued from previous page.

Code*	Title	EU	U.S.
c.5.2.2.5	Reprocessing	—	(Reprocessing)
c.5.2.2.6	Packaging Information	—	Packaging Information
c.5.2.2.7	GMP Statement	—	GMP Statement
c.5.2.3	Validation of the Process	II.B.3 Experimental data for the Validation of the method of manufacture	
c.5.3	Control tests on the finished product		
c.5.3.1	Specifications and routine tests	II.E.1 Specifications and routine tests	Specifications and analytical methods for drug product
c.5.3.1.1	Product specifications and tests for release (at time of manufacture) (General characteristics, specific standards)	II.E.1.1 Product specifications and tests for release	(Specifications)
c.5.3.1.2	Control methods	II.E.1.2 Control methods	(Monograph: Finished Product)
c.5.3.1.2.1	Test procedures for identification and quantitative determination for the active ingredient(s) must be described in detail (including biological and microbiological methods where relevant), together with other tests which include those in the appropriate general monograph for the type of dosage form in the European Pharmacopeia	II.E.1.2.1 Test procedures for identification and quantitative determination for the active substance(s)	
c.5.3.1.2.1.1	Identification tests	II.E.1.2.1 Identification tests	

*Key: "c" = Chemical, Pharmaceutical, and Biological Documentation

Table continued on next page.

S-15.01   page 18 of 45

Table continued from previous page.

Code*	Title	EU	U.S.
c.5.3.1.2.1.2	Quantitative determination of active substance(s)	II.E.1.2.1.- Quantitative determination of active substance(s)	
c.5.3.1.2.1.3	Purity tests	II.E.1.2.1.- Purity tests	
c.5.3.1.2.1.4	Pharmaceutical tests (e.g., dissolution)	II.E.1.2.1.- Pharmaceutical tests e.g., dissolution)	(Dissolution Test—General Method)
c.5.3.1.2.2	Identification and determination of excipients	II.E.1.2.2. Identification and determination of excipient(s)	
c.5.3.1.2.2.1	Identification tests for approved colouring materials	II.E.1.2.2. Identification tests for approved colouring materials	
c.5.3.1.2.2.2	Determination of antimicrobial or chemical preservatives (with limits)	II.E.1.2.2.- Determination of antimicrobial or chemical preservatives (with limits)	
c.5.3.2	Scientific data	II.E.2. Scientific data	
c.5.3.2.2	Batch analysis	II.E.2.2. Batch analysis	(Batch analysis)
c.5.3.2.2.1	Batches tested	II.E.2.2.- Batches tested	
c.5.3.2.2.2	Results obtained	II.E.2.2.- Results obtained	
c.5.3.2.2.3	Reference material (analytical results) primary and others	II.E.2.2.- Reference material (analytical results) primary and others	
c.5.4	Stability	—	—

Key: "c" = Chemical, Pharmaceutical, and Biological Documentation

Table continued on next page.

S-15.01   page 19 of 45

Table continued from previous page.

Code*	Title	EU	U.S.
c.5.4.1	Stability tests on active substance(s)	II.F.1 Stability tests on active substance(s)	(Stability of Active Constituents)
c.5.4.1.1	Batches tested	II.F.1 Batches tested	
c.5.4.1.2	General test methodology	II.F.1.- General test methodology	
c.5.4.1.2.1	Accelerated test conditions	II.F.1.-. Accelerated test conditions	
c.5.4.1.2.2	Normal test conditions	II.F.1.-. Normal test conditions	
c.5.4.1.3	Analytical test procedures	II.F.1.- Analytical test procedures	
c.5.4.1.3.1	Assay	II.F.1.-. Assay	
c.5.4.1.3.2	Determination of degradation products	II.F.1.-. Determination of degradation products	
c.5.4.1.5	Results of tests	II.F.1.- Results of tests	
c.5.4.1.6	Conclusions	II.F.1.- Conclusions	
c.5.4.2	Stability of Drug Product	II.F.2 Stability tests on the finished medicinal product	Stability of Drug Product
c.5.4.2.1	Quality Specification for the Proposed Shelf Life	II.F.2.- Quality Specification for the Proposed Shelf Life	
c.5.4.2.2	Batches tested and packaging	II.F.2.- Batches tested and packaging	
c.5.4.2.3	Study Methods	II.F.2.- Study methods	

Key: "c" = Chemical, Pharmaceutical, and Biological Documentation

Table continued on next page.

S-15.01

Table continued from previous page.

Code*	Title	EU	U.S.
c.5.4.2.3.1	Real Time Studies	II.F.2.-. Real Time Studies	
c.5.4.2.3.2	Studies under other Conditions	II.F.2.-. Studies under other conditions	
c.5.4.2.4.1	Characteristics studied	II.F.2.-. Characteristics studied	
c.5.4.2.4.1.1	Physical Characteristics	II.F.2.-. Physical Characteristics	
c.5.4.2.4.1.2	Microbiological Characteristics	II.F.2.-. Microbiological Characteristics	
c.5.4.2.4.1.3	Chemical Characteristics	II.F.2.-. Chemical Characteristics	
c.5.4.2.4.1.4	Biological Characteristics	—	
c.5.4.2.4.1.5	Chromatographic Characteristics	II.F.2.-. Chromatographic Characteristics	
c.5.4.2.4.5	Characteristics of the Packaging (Container/Closure Interaction with the Product)	II.F.2.-. Characteristics of the Packaging (Container/Closure Interaction with the Product)	
c.5.4.2.5	Evaluation Test Procedures	II.F.2.-. Evaluation Test Procedures	
c.5.4.2.5.1	Description of Test Procedures	II.F.2.-. Description of Test Procedures	Stability indicating Methods
c.5.4.2.6	Results of Test	II.F.2.-. Results of Test	
c.5.4.2.6.1	Stability Data	—	Stability Data
c.5.4.2.6.2	Supporting Stability Data		Supporting Stability Data

Key: "c" = Chemical, Pharmaceutical, and Biological Documentation

Table continued on next page.

S-15.01   page 21 of 45

Table continued from previous page.

Code*	Title	EU	U.S.
c.5.4.2.7	Conclusions	II.F.2.- Conclusions	Summary and interpretation of stability
c.5.4.2.7.1	Shelf life and storage conditions	II.F.2.-. Shelf life and storage conditions	(Recommended expiration date—shelf life)
c.5.4.2.7.2	Shelf life after Reconstitution and/or first Opening of the Product	II.F.2.-. Shelf life after Reconstitution and/or first Opening of the Product	
c.5.4.2.7.3	Stability and compatibility with accessories used for the administration of the product		
c.5.4.2.8	Ongoing Stability Studies	II.F.2.- Ongoing Stability Studies	
c.6	Environmental Assessment	III.R Environmental Risk Assessment II.H Data related to the environmental risk assessment for products containing/consisting of genetically modified organisms (GMOs)	d.1.iii Environmental Impact Analysis Report
c.7	Validation	—	—
c.7.1	Method Validation Package	—	Method Validation Package
c.7.1.1	Table of contents (NDA)	—	Table of contents
c.7.1.2	Samples for Validation	—	
c.7.1.2.1	List of samples for validation	—	List of samples for validation

Key: "c" = Chemical, Pharmaceutical, and Biological Documentation

Table continued on next page.

S-15.01   page 22 of 45

Table continued from previous page.

Code*	Title	EU	U.S.
c.7.1.2.2	Identity of samples for validation	—	Identity of samples for validation
c.7.1.3	Test methods and specifications	—	Test methods and specifications
c.7.1.3.1 1 of 3	Drug Substance	II.C.1.2.5.(A)- Analytical validation and comments on the choice of routine tests and standards	Drug Substance
c.7.1.3.1 2 of 3	Drug Substance (for herbal medicines)	II.C.1.2. 5.(C)1.- Analytical development and validation, commentary on the choice of routine tests and specifications	
c.7.1.3.1 3 of 3	Drug Substance	II.C.1.2. 5.(C)2.- Analytical development and validation, commentary on the choice of routine tests and specifications	
c.7.1.3.2	Drug Product	II.E.2.1.- Analytical validation and comments on the choice of routine tests and standards	Drug Product
c.7.1.3.3	Biological	—	
c.7.1.3.3.1	Validation of the relevant methods used during the development	—	
c.7.1.3.3.2	Validation and comments on the choice of routine tests	—	

Key: "c" = Chemical, Pharmaceutical, and Biological Documentation

Table continued on next page.

Table continued from previous page.

Code*	Title	EU	U.S.
c.7.1.3.3.3	Characterization of the reference material	—	
c.7.1.3.4	Stability Methods validation		
c.7.1.3.4.1	Drug Substance	II.F.1.- Validation of all test procedures including limits of detection (including initial results)	
c.7.1.3.4.2	Drug Product	II.F.2.- Validation of test procedures	(Suitability of analytical methods)
c.7.1.4	Test results	—	Test results
c.7.2	Process validation	—	
c.7.2.1	Validation of the production process, data on consistency of the yield and degree of purity and on the quality of the active ingredient	—	
c.7.2.2	Removal of impurities during the purification process	—	
c.7.2.3	Lifetime of purification columns	—	
c.7.2.4	Stability of any intermediate of production and/or manufacturing when intermediate storage are intended in the process	—	
c.7.2.5	In the event of any reprocessing being necessary, full validation of each proposed step	—	

*Key: "c" = Chemical, Pharmaceutical, and Biological Documentation

Table continued on next page.

S-15.01　page 24 of 45

Table continued from previous page.

Code*	Title	EU	U.S.
c.8	Other information	—	—
c.8.1	Other	II.Q Other	—
d	*Samples/labeling*		
d.1	Registration Samples	—	—
d.1.1	Samples on Request	—	e.1 Samples on Request
d.1.1.1	4 samples each, sufficient for 3x every Test described	—	e.1.i 4 samples each, sufficient for 3x every Test described
d.1.1.1.1	Finished Product	—	e.1.i.a Drug Product
d.1.1.1.2	Drug Substance used in the Samples under d.1.1.1.1	—	e.1.i.b Drug Substance used in the Samples under (a)
d.1.1.1.3	Reference Standards, Blanks (except recognized Pharmacopoeial Standards)	—	e.1.i.c Reference Standards, Blanks (except recognized Pharmacopeial Standards)
d.1.2	Samples	I.A Samples	—
d.1.2.1	List of Samples	List of Samples	—
d.2	International Physicians Circular	—	
d.2.1	Pharmacological Class/Rational Use/Benefits	—	c.2.ii Pharmacological Class/Rational Use/Clinical Benefits
d.2.1.1.2	Samples of the Finished Market Package	—	e.1.ii Samples of the Finished Market Package on request

*Key: "c" = Chemical, Pharmaceutical, and Biological Documentation; "d" = Samples/Labeling

Table continued on next page.

Table continued from previous page.

Code*	Title	EU	U.S.
d.2.2	Summary of Product Characteristics	I.B.1 Summary of Product Characteristics	
d.3	Label and Carton Text	—	
d.3.1 1 of 2	Labeling Text with Annotations	—	c.2.i, e.2.ii Labeling Text with Annotations
d.3.1 2 of 2	Labeling Text with Annotations	—	e.2.ii Copies of label and all labeling for drug product
d.3.11	Labeling Text without Annotations		Nonannotated Package Insert
d.3.2	Packaging	I.B.2.- Packaging	
d.3.3	Labeling	I.B.2.- Labeling	
d.4	Declaration		
d.4.2	Patient Package Insert		—
d.4.2.2	Package Insert	I.B.2. Package Insert(s)	
d.5	Other		—
d.5.1	Manufacturers Authorization(s)	I.A Manufacturers Authorization(s)	
e	***Nonclinical/Tox Pharmacology***	—	—
e.1	Table of Contents	III Table of Contents	—
e.1.2	Table of Contents EU	III Table of Contents	—
e.3	Pharmacology Studies	III.F Pharmacodynamics	d.2.ii Pharmacology Studies

*Key: "d" = Samples/Labeling; "e" = Nonclinical/Tox Pharmacology

Table continued on next page.

S-15.01

Table continued from previous page.

Code*	Title	EU	U.S.
e.3.1	Specific nonclinical pharmacology studies	III.F.1 Pharmacodynamic effects relating to proposed indications	Studies of the pharmacological actions of the drug in relation to proposed therapeutic indications
e.3.2	Studies that otherwise define pharmacologic properties or are pertinent to possible adverse effects	—	Studies that otherwise define pharmacologic properties or are pertinent to possible adverse effects
e.3.2.1	General nonclinical pharmacology studies	III.F.2 General Pharmacodynamics	
e.3.2.2	Animal drug interaction studies	III.F.3 Drug interactions	
e.4	Toxicity Studies	—	
e.4.1	Organ Toxicity	—	
e.4.1.1	Single dose studies	—	
e.4.1.1.1	Acute toxicity studies	—	Acute toxicity studies
e.4.1.1.1.1	Acute toxicity rodent	Single dose toxicity	at least 2
e.4.1.1.1.2	Acute toxicity nonrodent	—	at least 1, also range-finding study may be substituted
e.4.1.2	Repeated dose studies	—	—
e.4.1.2.1.1	Subacute toxicity studies (≤ 3 months)	Subacute toxicity trials (up to 3 months)	subacute toxicity studies
e.4.1.2.1.1.1	Subacute toxicity studies (≤ 14 days)	1 or several doses in 1d in man: test 2 wks in animal	

*Key: "e" = Nonclinical/Tox Pharmacology

Table continued on next page.

Table continued from previous page.

Code*	Title	EU	U.S.
e.4.1.2.1.1.1.1	Rodent	for subacute testing at least 1 rodent	
e.4.1.2.1.1.1.2	Nonrodent	for subacute testing at least 1 nonrodent	
e.4.1.2.1.1.2	> 14 days ≤ 1 month	for administration in man up to 7d test 4 wks in animal (generally 1 test 2–4 wks required)	
e.4.1.2.1.1.2.1	Rodent	for subacute testing at least 1 rodent	
e.4.1.2.1.1.2.2	Nonrodent	for subacute toxicity testing at least 1 nonrodent	
e.4.1.2.1.1.3	Subacute toxicity studies > 1 month ≤ 3 months	for use in man up to 30d test 3m in animal	for up to 2 wks use in man, test for 2 wks to 3m in animal
e.4.1.2.1.1.3.1	Rodent		for subacute tests usually rat
e.4.1.2.1.1.3.2	Nonrodent		for subacute testing usually dog; primate may be chosen if PD or PK more suitable
e.4.1.2.1.2	Subacute dose-range finding	—	
e.4.1.2.1.2.1	Rodent	—	
e.4.1.2.1.2.2	Nonrodent	—	
e.4.1.2.2	Chronic toxicity studies (> 3 months)	IIIB2. chronic toxicity trials (beyond 3 months)	for use in man up to 3m test 6m in animal

*Key: "e" = Nonclinical/Tox Pharmacology

Table continued on next page.

S-15.01   page 28 of 45

Table continued from previous page.

Code*	Title	EU	U.S.
e.4.1.2.2.1	Rodent	usually 1 rodent	for chronic testing usually rat
e.4.1.2.2.2	Nonrodent	at least 1 nonrodent	for chronic toxicity testing usually dog
e.4.2	Special toxicity	—	—
e.4.2.1	Animal local tolerance/Toxicity studies	III.H Local tolerance	. . . studies of toxicities related to the drug's particular mode of administration or conditions of use
e.4.2.1.1	Local tolerance	e.g., for topicals	
e.4.2.1.1.1.1	Skin		
e.4.2.1.1.1.2	Mucosa		
e.4.2.1.1.1.3	Parenteral application (i.v., i.a., s.c., i.m., paravenous) (1x)		
e.4.2.1.2	Phototoxicity		phototoxic properties may be evaluated on rats, mice, and rabbits
e.4.2.1.3	Sensitization	possibility of sensitization shall be investigated	sensitization
e.4.2.1.3.2	Skin		
e.4.2.1.3.3	Respiratory Tract		
e.4.2.1.3.4	Photosensitization		photoallergic effects may be evaluated in man

Key: "e" = Nonclinical/Tox Pharmacology

Table continued on next page.

S-15.01   page 29 of 45

Table continued from previous page.

Code*	Title	EU	U.S.
e.4.2.4	Immunotoxicity	immunotoxicity (asked for in the expert report)	
e.4.2.4.1	Antigenicity		
e.4.2.4.2	Cross-reactivity		
e.4.2.5	Reproduction studies	—	
e.4.2.5.0.1	Dose range-finding		
e.4.2.5.0.2	Rearing		
e.4.2.5.1	Reproduction Segment I: Fertility and general reproductive performance studies	III.B Fertility and general reproductive performance	studies as appropriate, of the effects of the drug on reproduction and on the developing fetus
e.4.2.5.1.1	Rodent	at least 1 species	1 species, rats most frequently used (mouse can be used instead)
e.4.2.5.1.2	Nonrodent	—	
e.4.2.5.2	Reproduction Segment II: Embryotoxicity (particularly teratogenicity) studies	III.C Embryofoetal and perinatal toxicity	d.2.iii studies as appropriate, of the effects of the drug on reproduction and on the developing fetus
e.4.2.5.2.1	Rodent	1 species required, usually rat or mouse	at least 2 species required from rat, mouse (and rabbit)
e.4.2.5.2.2	Nonrodent	1 species required (rabbit)	at least 2 species required (rat, mouse) or rabbit. Other species such as dogs, cats, pigs, etc. have been used

Key: "e" = Nonclinical/Tox Pharmacology

Table continued on next page.

S-15.01   page 30 of 45

Table continued from previous page.

Code*	Title	EU	U.S.
e.4.2.5.3	Reproduction Segment III: peri/postnatal toxicity studies	Peri/postnatal toxicity	peri/postnatal toxicity
e.4.2.5.3.1	Rodent	at least 1 species required	1 species, usually rat
e.4.2.5.4	Multigeneration studies	—	
e.4.2.5.4.1	Rodent	—	
e.4.2.6	Oncogenicity/Carcinogenicity studies	—	—
e.4.2.6.1	Preliminary carcinogenicity studies (short-term tests)	—	—
e.4.2.6.1.1	Acute toxicity (short-term)	—	—
e.4.2.6.1.1.1	Rodent	—	—
e.4.2.6.1.2	Subacute toxicity	—	—
e.4.2.6.1.2.1	Rodent	—	—
e.4.2.6.2	Carcinogenicity studies	III.E Carcinogenic potential	Carcinogenicity studies
e.4.2.6.2.1	Rodent	2 species required, usually rodents	2 species required, usually rat and mouse
e.4.2.6.2.2	Nonrodent	—	Dog in case of oral contraceptives
e.4.2.7	Genotoxicity	—	—
e.4.2.7.1	Genotoxicity	—	—
e.4.2.7.1.1	Test for point mutation	III.D Mutagenic potential	—

*Key: "e" = Nonclinical/Tox Pharmacology

Table continued on next page.

S-15.01   page 31 of 45

Table continued from previous page.

Code*	Title	EU	U.S.
e.4.2.7.1.1.1	In vitro tests for point mutation	III.D.1 In vitro	
e.4.2.7.1.1.1.1	Ames Test	—	
e.4.2.7.1.1.1.2	HGPRT test		
e.4.2.7.1.1.1.3	Other in vitro tests		
e.4.2.7.1.1.2	In vivo tests for point mutations	III.D.2 In vivo	
e.4.2.7.1.2	Test for chromosomal effects		
e.4.2.7.1.2.1	In vitro tests for chromosomal aberrations		
e.4.2.7.1.2.2	In vivo		
e.4.2.7.1.2.2.1	Micronucleus test		
e.4.2.7.1.2.2.2	Other		
e.4.2.7.1.3	Test for DNA effects		
e.4.2.7.1.3.1	In vitro		
e.4.2.7.1.3.1.1	UDS test		
e.4.2.7.1.3.1.2	Other tests		
e.4.2.7.1.3.2	In vivo		
e.4.2.7.1.3.2.1	UDS test		
e.4.2.7.1.3.2.2	Other tests		

Key: "e" = Nonclinical/Tox Pharmacology

Table continued on next page.

S-15.01   page 32 of 45

Table continued from previous page.

Code*	Title	EU	U.S.
e.4.2.7.2	Other tests/indicator tests		
e.4.2.7.2.1	In vitro		
e.4.2.7.2.2	In vivo		
e.5	Animal-Pharmacokinetics/ADME	—	Animal-Pharmacokinetics/ADME
e.5.1	Pharmacokinetics in animals after single dose	III.G.1 Pharmacokinetics after a single dose	any studies of the absorption, distribution, metabolism, and excretion of the drug in animals
e.5.1.1	PK after single dosing	—	
e.5.1.1.1	Rodent	—	
e.5.1.1.1.1	Rat	—	
e.5.1.1.1.2	Rodent other than Rat	—	
e.5.1.1.2	Nonrodent	—	
e.5.1.1.2.1	Second species (nonrodent)	—	
e.5.1.1.2.2	Additional Species (nonrodent)	—	
e.5.1.2.1	Toxicokinetics (T)	—	
e.5.1.2.2	Toxicokinetics (PK)	—	
e.5.1.2.2.1	Rodent	—	
e.5.1.2.2.1.1	Rat	—	

Key: "e" = Nonclinical/Tox Pharmacology

Table continued on next page.

Table continued from previous page.

Code*	Title	EU	U.S.
e.5.1.2.2.1.2	Rodent other than Rat	—	
e.5.1.2.2.2	Nonrodent	—	
e.5.1.2.2.2.1	Usual species	—	
e.5.1.2.2.2.2	Additional Species	—	
e.5.2	Pharmacokinetics in animals after repeated dose	III.G.2 Pharmacokinetics after repeated administration	any studies of the absorption, distribution, metabolism, and excretion of the drug in animals
e.5.2.1	Rodent	—	
e.5.2.1.1	Rat	—	
e.5.2.1.2	Rodent other than Rat	—	
e.5.2.2	Nonrodent	—	
e.5.3	Distribution in normal and pregnant animals	III.G.3 Distribution in normal and pregnant animals	d.2.iv any studies of the absorption. distribution, metabolism, and excretion of the drug in animals
e.5.3.1	Autoradiography in male, female, and pigmented rats	—	
e.5.3.2	Quantitative distribution study in male rats	—	
e.5.3.3	Autoradiography in pregnant rats	—	
e.5.3.4	Placental transfer (quantitative)	—	

*Key: "e" = NonClinical/Tox Pharmacology

Table continued on next page.

S-15.01   page 34 of 45

Table continued from previous page.

Code*	Title	EU	U.S.
e.5.3.5	Protein binding	—	
e.5.3.5.1	Protein binding in different species in vitro	—	
e.5.3.5.2	Binding characteristics and interactions in vitro	—	
e.5.4	Biotransformation	—	—
e.5.4.1	In animals	III.G.4 Biotransformation	any studies of the absorption, distribution, metabolism, and excretion of the drug in animals
e.5.4.1.1	Rodent	—	
e.5.4.1.1.1	Rat	—	
e.5.4.1.1.2	Rodent other than Rat	—	
e.5.4.1.2	Nonrodent	—	
e.5.4.1.2.1	Usual Species	—	
e.5.4.1.2.2	Additional Species	—	
e.5.4.2	In vitro	—	—
e.6	GLP Compliance Statement	in the expert report, the expert should comment on the GLP status of the studies submitted	d.2.v for each nonclinical laboratory study subject to the GLP regulations, a statement that it was conducted in compliance with those regulations or a brief statement of the reason for the noncompliance

*Key: "e" = Nonclinical/Tox Pharmacology

Table continued on next page.

S-15.01

Table continued from previous page.

Code*	Title	EU	U.S.
e.7	Other Information on Toxicology, Pharmacology, and/or Kinetics	III.Q Other information	
e.7.1	Enzyme induction		
e.7.2	Test on covalent binding (protein, DNA)		
e.7.3	Impurities		
e.7.4	Other (e.g., compatibility with blood)		
f	**Human Pharmacokinetics and Bioavailability Data**	—	—
f.2	Description, Compliance Statements	—	d.3.i description of each of the bioavailability and pharmacokinetic studies of the drug in humans performed by or on behalf of the applicant that includes a description of the analytical and statistical methods
f.3	Rationale of Methods, Specifications of Drug Product with Regard to Bioavailability	—	d.3.ii if the applicat. describes in the CMC sect. specifications or analyt. meth. needed to assure the bioavail. of the drug prod. or subst. or both, a statement in this sect. of the rationale for establishing the specific. or analyt. meth., incl. data and information supporting the rationale
f.3.2	Abbreviations	—	
f.3.3	Precis	—	

*Key: "e" = Nonclinical/Tox Pharmacology; "f" = Human Pharmacokinetics and Bioavailability Data

Table continued on next page.

Table continued from previous page.

Code*	Title	EU	U.S.
f.3.4	Chemistry	—	
f.3.4.1	Physicochemical properties	—	
f.3.4.2	Dosage forms	—	
f.3.5	Analytical Methods in Biological Fluids		
f.3.6	In vitro Stability in Biological Fluids	—	
f.3.7	Structure of Drugs with radiolabeled position indicated		
f.3.8	Protein Binding	—	
f.4	Clinical Pharmacokinetics, Bioavailability & ADME studies	IV.A.2 Pharmacokinetics	Human Pharmacokinetics and Bioavailability Section
f.4.1	Studies concerning absorption		
f.4.2	Studies concerning distribution/bioavailability	II.G and IV.Q.1 Bioavailability/ Bioequivalence	—
f.4.3	Studies concerning metabolism	II.Q Studies concerning metabolism	
f.4.4	Studies concerning elimination		
f.5	Comparison of Pharmacokinetics in Animals and Man	—	

*Key: "f" = Human Pharmacokinetics and Bioavailability Data

Table continued on next page.

S-15.01   page 37 of 45

Table continued from previous page.

Code*	Title	EU	U.S.
f.6	Summarizing discussion	—	d.3.iii summarizing discussion and analysis of the pharmacokinetics and metabolism of the active ingredient(s) and the bioavailability or bioequivalence, or both, of the drug product
g	**Microbiology Data**	—	
g.1	Table of Contents	—	
g.1.1	Table of Contents NDA	—	d.4 Microbiology section
g.2	Biochemical Basis		d.4.i Description of the biochemical basis of the drug's actions on microbial physiology
g.2.1	Mechanism of action		
g.2.1.1	Mode of action, general		
g.2.1.2	Chemical structure		
g.2.1.3	Relation to other (related) compounds		
g.2.1.4	Relation to other antibiotics of the same group		
g.3	Antimicrobial spectra incl. in vitro preclinical studies		d.4.ii description of the antimicrobial spectra of the drug, including results of in vitro preclinical to demonstrate concentrations of the drug required for effective use
g.3.1	Antimicrobial Activity		

*Key: "*P*" = Human Pharmacokinetics and Bioavailability Data; "*g*" = Microbiology Data

Table continued on next page.

Table continued from previous page.

Code*	Title	EU	U.S.
g.3.1.1	General		
g.3.2	Enzyme Hydrolysis Rates		
g.3.3	Miscellaneous Studies		
g.3.3.1	Effect of Medium, pH, and Test Conditions		
g.3.3.2	Inoculum effects		
g.3.3.3	Binding proteins		
g.3.3.4	Permeability		
g.4	Mechanisms/Assessment of resistance		d.4.iii Description of any known mechanisms of resistance to the drug including results of any known epidemiologic studies to demonstrate prevalence of resistance factors
g.4.1	Activity against Clinical Trial isolates resistant to other antibiotics		
g.4.2	Induced resistance		
g.5	Clinical microbiology methods needed for effective use of drug		d.4.iv description of clinical microbiology laboratory methods (for example in vitro sensitivity discs) needed for effective use of drug
g.5.1	Clinical Laboratory Susceptibility Test Methods		
g.5.11	Class concept		

*Key: "g" = Microbiology Data

Table continued on next page.

S-15.01   page 39 of 45

Table continued from previous page.

Code*	Title	EU	U.S.
g.5.1.2	Dilution methods		
g.5.1.3	Disk diffusion methods		
g.5.2	In Vivo Animal Protection Studies	—	
g.5.3	In Vitro Studies Conducted During Clinical Trials	—	
h	**Clinical Data**	—	
h.1	Table of Contents	—	
h.1.2	Table of Contents EU	IV Table of Contents	
h.2	Clinical Pharmacology (incl. pharmacokinetics)	IV.A Clinical pharmacology	d.5.i Clinical pharmacology studies
h.2.1	Pharmacodynamics	IV.A.1 Pharmacodynamics	
h.3	Clinical Experience	IV.B Clinical experience	
h.3.1	Clinical trials	IV.B.1 Clinical trials	
h.3.1.1 1 of 2	Controlled Studies		d.5.ii Controlled clinical studies
h.3.1.1.1	Placebo-controlled studies	Placebo-controlled studies	
h.3.1.1.1.1	Placebo		—
h.3.1.1.1.1.1	Full report		—
h.3.1.1.1.2	Short report		—

Key: "g" = Microbiology Data; "h" = Clinical Data

Table continued on next page.

S-15.01   page 40 of 45

Table continued from previous page.

Code*	Title	EU	U.S.
h.3.1.1.1.2	Placebo/add-on-therapy	—	
h.3.1.1.1.2.1	Full report	—	
h.3.1.1.1.2.2	Short report	—	
h.3.1.1.2	Controlled studies with reference therapies	Controlled studies with reference therapies	
h.3.1.2	Uncontrolled studies	Noncontrolled studies	d.5.iii Uncontrolled clinical studies
h.3.1.2.1	Final report	—	
h.3.1.2.2	Interim report		
h.3.1.3	Other studies and information	IV.B.3 Published and unpublished experience	d.5.iv Other Studies and Information
h.3.1.3.0.1	Pilot studies in patients	—	
h.3.1.3.0.2	Dose finding studies	—	
h.3.1.3.0.3	Long term safety	—	
h.3.1.3.0.4	Special patient groups	—	
h.3.1.3.0.5	Dose/effect relationship	—	
h.3.1.3.0.6	Special topics (e.g., renal insuff.)	—	

*Key: "h" = Clinical Data

Table continued on next page.

S-15.01   page 41 of 45

Table continued from previous page.

Code*	Title	EU	U.S.
h.3.1.3.1	Brief information on ongoing trials and uncompleted trials (incl. the reason why the trials were not completed) with full details on any safety issues raised in these studies	IV.B.3.1 Brief information on ongoing trials and uncompleted trials	Controlled clinical studies that have not been analyzed in detail for any reason (e.g., because they have been discontinued or are incomplete) are to be included in this section including a copy of the protocol and a brief description of the results and status of the study
h.3.1.3.2	Any other information	IV.B.3.2 Any other information	
h.4	Integrated summary (effectiveness)	—	d.5.v integrated summary (effectiveness)
h.5	Integrated summary (safety)	—	d.5.vi Summary and update of safety information
h.5.1 1 of 2	Integrated summary (safety)	—	d.5.vi.a Integrated summary (safety)
h.5.1 2 of 2	Copy of information submitted under clinical section, integrated summary safety	—	d.6.ii Copy of information submitted under clinical section, integrated summary safety [= d.5.vi.a]
h.6	Potential of abuse	—	d.5.vii Potential of abuse, drug abuse/overdose information
h.7	Integrated summary of benefits/risks	—	d.5.viii Integrated summary of benefits/risks
h.8	Technical	—	—
h.8.1	Compliance statements for studies in humans	—	d.5.ix Compliance statements for studies in humans

*Key: "h" = Clinical Data

Table continued on next page.

Table continued from previous page.

Code*	Title	EU	U.S.
h.8.2	Name and address of CROs, obligations transferred	—	d.5.x Name and address of CROs, obligations transferred
h.8.3	Studies, where original subject records were audited or reviewed by the sponsor	—	d.5.xi List identifying each clinical study where original subject records were audited or reviewed by the sponsor in the course of monitoring to verify the accuracy
h.9	Other clinical information	IV.Q Other information	—
j	***Safety and other Update Reports***	—	—
j1	Safety update	—	d.5.vi.b Safety update
j.1.1	Postmarketing experience	IV.B.2 Postmarketing experience	
j.1.1.1	Adverse reactions and monitoring events and reports	IV.B.2.1 Adverse reactions and monitoring events and reports	
j.1.1.2	Number of patients exposed	IV.B.2.2 Number of patients exposed	
j.2	Updating Marketing Authorization Data	—	—
j.2.1	Report on experience (2 years) German Drug Law § 49	—	—
j.2.2	Report on experience (5 years)	—	—
j.2.3	Report on assessment criteria alteration for prolongation of marketing authorization (each 5 years) German Drug Law § 31	—	—

Key: "h" = Clinical Data; "j" Safety and other Update Reports

Table continued on next page.

S-15.01   page 43 of 45

Table continued from previous page.

Code*	Title	EU	U.S.
k	**Statistical Data/Overall Summary**	—	**Statistical Data/Overall Summary**
l	**Cover sheet(s)**	—	—
le	**Cover sheet(s) EU**	—	—
le.1	Cover sheet(s)	I Summary of the dossier	—
le.2.1	Cover sheet(s)	II Chemical, pharmaceutical, and biological documentation	—
le.3.1	Cover sheet(s)	III Toxicological and pharmacological documentation	—
le.4.1	Cover sheet(s)	IV Clinical Documentation	—
lu	**Cover sheet(s) U.S**	—	
lu	Cover sheet(s)	—	a Application form
lu	Cover sheet(s)	—	d.2 Nonclinical Pharmacology/Toxicology Section
lu	Cover sheet(s)	—	e Samples/Labeling
lu	Cover sheet(s)	—	Batch Information
lu	Cover sheet(s)	—	dTechnical Sections
lu	Cover sheet(s)	—	c Summary
lu	Cover sheet(s)	—	d.1 Chemistry/Manufacturing and Controls section

Table continued on next page.

S-15.01   page 44 of 45

Key: "k" = Statistical Data/Overall Summary; "l" = Cover sheet(s); "le" = Cover sheet(s) EU; "lu" = Cover sheet(s) U.S.

Table continued from previous page.

Code*	Title	EU	U.S.
lu	Cover sheet(s)	—	d.1.ii Drug Product
lu	Cover sheet(s)	—	f Case Report Forms & Tabulations
lu	Cover sheet(s)	—	d.1.i Drug Substance
lu	Cover sheet(s)	—	d.5 Clinical Data Section
lu	Cover sheet(s)	—	Summary
lu	Cover sheet(s)	—	d.2.ii Toxicology Studies
m	*Raw Data*	—	
m.1.1	Case Report Tabulations	—	f.1 Case report tabulations
m.1.2	Case Report Forms	—	f.2 Case report forms
m.1.3	Additional Data	—	f.3 Additional data

*Key: "lu" = Cover sheet(s) U.S.; "m" = Raw Data

A Department XYZ Standard

20.01: Labeling

Document Type:	Standard
Document Code:	20.01
	(enter company-specific code)
Title:	Labeling
Date/Revision No.:	DD/MM/YY number xy
Scope:	Global
References:	(enter policies, standards, SOPs of your department/company, or other documents [e.g., guidelines] that should be considered in this context)
	implements Policy on Labeling (P-20)
Authorization:	
	Signature of authorized person(s)
	Name of authorized person(s)
	Job title/Function of authorized person(s)
Issue Date:	DD/MM/YY
Implementation Date:	DD/MM/YY

Labeling must contain the following information:

- Pharmaceutical and therapeutic category
- Prescription status
- Active ingredient(s)
- Other excipients with pharmacodynamic or medicinal significance
- Other excipients
- Storage and stability
- Storage and stability after opening of the container
- Information on handling, preparation of final dosage form, opening, measurements of doses
- Mode of action
- Toxicological information
- Pharmacodynamic properties
- Pharmacokinetic properties
- Interaction with other medicinal products and other forms of interaction (e.g., caffeine, alcohol, nicotine, food)
- Indications
- Posology and method of administration
- Contraindications
- Special warnings and precautions for use
- Undesirable effects
- Use in women with child-bearing potential, use during pregnancy and lactation
- Tolerance
- Dependence
- Off-label use
- Overdose
- Interference with laboratory tests
- Warnings

A Department XYZ Standard

27.01: Submission

Document Type:	Standard
Document Code:	27.01
	(enter company-specific code)
Title:	Submission
Date/Revision No.:	DD/MM/YY number xy
Scope:	Global
References:	(enter policies, standards, SOPs of your department/company, or other documents [e.g., guidelines] that should be considered in this context)
	1. Policy on Submission (P-27)
	2. Department operational procedures
Authorization:	
	Signature of authorized person(s)
	Name of authorized person(s)
	Job title/Function of authorized person(s)
Issue Date:	DD/MM/YY
Implementation Date:	DD/MM/YY

1. PURPOSE

In order to meet the rules as set out in the policy on submissions (1), quality is defined by this standard through specifying quality of input and output and identifying key processes to be vali dated.

2. DEFINITIONS

The key terms pertaining to this policy should be defined here. As there are no uniform and globally accepted definitions available, please develop your own definitions. In this way, the language of the staff of your organization can be incorporated.

- The term *dossier* signifies a compilation of documents for a specific regulatory purpose (e.g., application for clinical trial authorization or application for marketing authorization) in a specified country(ies) for a developmental or already marketed medicinal product in a structured form (i.e., submission-like). If applicable, it is a subset of the global dossier. The dossier is the basis for the submission(s).

- The term *global dossier* signifies a compilation of all documents required for international regulatory purpose(s) for a developmental or already marketed medicinal product. It is maintained continuously throughout the life cycle of the medicinal product and serves as a repository for the generation of dossiers and submissions.

- The term *submission* signifies a country-specific compilation of documents for a specific regulatory purpose (e.g., application for clinical trial authorization or application for marketing authorization) for a developmental or already marketed medicinal product in a structured form according to national regulatory requirements. It is based on the dossier, or, if applicable, the global dossier. It may contain additional national documents (e.g., national leaflets or application forms).

3. STATEMENT OF STANDARD

1. Quality of Processes

The following processes (if applicable) are adequately standardized and validated (2): Generation of Table of Contents (TOC), paginating master copy, copying from master copy to produce copies for Regulatory Bodies, putting in binders, labeling, export to MO or CD, mailing submission.

2. Quality of Input

Specs/Limits	Frequency/ Extent of Checks	Documentation of Check Results	Responsibility for Checks	Duplication of Checks by RA
1. Original national docs				
- Complete (all pages)	every doc: every p	monthly report to head RA	RA	—
- Format and content acc. to state of the art, SOPs	all items			
- Good copying quality	every p			
- Cross-references OK	all cross-references			
- Good English	main part			
2. Material				
2.1 Paper: Good copying quality	each delivery/1 p	yearly report to head RA	Copying department	None
2.2 Binders: stable, company Logo	each delivery/1 binder	yearly report to head RA	person responsible for ordering material	national RA: spot checks on master copy
2.3 Labels: good adhesion	each delivery/1 label	yearly report to head RA	person responsible for ordering material	check on master copy

Table continued on next page.

3. Quality of Output (= Submission)

Specs/Limits	Frequency/ Extent of Checks	Documentation of Check Results	Responsibility for Checks	Duplication of Checks by RA
1. Submission master copy		final report to head RA	national RA	—
- Correct sequence acc. to TOC	all docs/TOC			
- For expert reports, summaries: cross-references to documentation OK	all cross-references in expert reports, summaries			
2. Copies (from master copy)	spot check (1 copy or 10% of copies, whichever is greatest, summary part all pages, 1st vol of other parts all pages)	regular report to head RA	Copying department	National RA: spot check: 1 copy: summary part and 1 vol of another part all pages
- Complete (all pages)				
- Good copying quality				

A Department XYZ Standard

28.01: Terminology

Document Type:	Standard
Document Code:	28.01
	(enter company-specific code)
Title:	Terminology
Date/Revision No.:	DD/MM/YY number xy
Scope:	Global
References:	(enter policies, standards, SOPs of your department/company, or other documents [e.g., guidelines] that should be considered in this context)
	implements Policy on Terminology (P-28)
Authorization:	
	Signature of authorized person(s)
	Name of authorized person(s)
	Job title/Function of authorized person(s)
Issue Date:	DD/MM/YY
Implementation Date:	DD/MM/YY

Abbreviated New Drug Application (U.S.)	ANDA
Abbreviated New Drug Submission (Canada)	ANDS
Adverse Drug Event	ADE
Adverse Drug Reaction	ADR
Adverse Drug Reaction On-Line Information Trackingsystem (UK)	ADROIT
Advertising (see Promotion)	
Analysis of variance	ANOVA
Anatomical Therapeutic Chemical Classification Index	ATC Code
Application for a Clinical Trial License	A notification or application to a Regulatory Body with the purpose of starting local trials in humans.
Application for a Marketing Authorization	A notification of an application to a Regulatory Body with the purpose of placing a medicinal product on the market.
Archiving Management	The identification, retention, storage, protection, and disposal of the records of Regulatory Affairs, including records on the developmental research of medicinal products and already marketed medicinal products.
Arzneimittelgesetz (Germany) (Drug Law)	AMG
Autorisation de Mise Sur Marché (France) (marketing authorization)	AMM
British Approved Name	BAN
British Institute of Regulatory Affairs	BIRA
British Pharmacopoeia	BP
Case Record Form	CRF

Center for Biologics Evaluation and Research (U.S.)	CBER
Center for Drug Evaluation and Research (U.S.)	CDER
Central European Society of Regulatory Affairs	MEGRA
Central Pharmaceutical Affairs Council (Japan)	CPAC
Change	Any changes or variations concerning the company's medicinal products worldwide. This includes, for example, changes in starting materials, manufacturing process/equipment/batch size/site, dosage form, packaging, controls, patient leaflet, labeling, or outer packaging. If subject to submission to the Regulatory Bodies, any additional information, such as advertising/promotional material, is also included.
Chemical Abstracts Registry Number	CARN
Chemistry, Manufacturing, and Controls	CMC
Clinical Quality Assurance	CQA
Clinical Trial Application (UK)	CTA
Clinical Trial Certificate (UK)	The term signifies an approval by a Regulatory Body with regard to starting local trials in humans. CTC
Clinical Trial Exemption Application (UK)	CTX
Code of Federal Regulations (U.S.)	CFR
Coding Symbols for a Thesaurus of Adverse Reaction Terms	COSTART

Committée Européen de Normalisation	CEN
Committee for Proprietary Medicinal Products	CPMP
Committee for Veterinary Medicinal Products	CVMP
Company Comment	A document communicating to Regulatory Bodies comments—suggestions for changing a guideline. Typically, it will be in letter format and contain comments, suggestions, and critical issues, with references to the original document. It may be submitted via industry associations and/or Regulatory Affairs professional societies or directly to the Regulatory Bodies.
Company Position	For each important guideline the result of interpretation/discussion by Regulatory Affairs, and, if applicable, other concerned departments/disciplines. It will contain a summary and an evaluation of, for example, critical issues, possible consequences for the company, and recommendations.
Computer-Assisted Marketing Authorization	CAMA
Computer-Assisted New Drug Application	CANDA
Computer-Assisted Product License Application	CAPLA
Computer-Assisted Regulatory Submission	CARS
Contact Partner	The term signifies person(s), department(s), company(ies), and/or Regulatory Bodies with which Regulatory Affairs personnel interact in a business environment.

Contact Report	The term signifies a report on contact with Regulatory Bodies in a standardized format.
Contact with Regulatory Body	The term signifies any communication between Regulatory Affairs and a Regulatory Body directly by phone, E-mail, fax, telex, or letter, or during a personal meeting.
Contract Research Organization	CRO
Council for the International Organization of Medicinal Science	CIOMS
Crisis	The term signifies any situation with a high potential of danger or damage to the company's reputation, substances, or medicinal products, or personnel within the scope of Regulatory Affairs responsibilities or closely connected with its function or directly affecting Regulatory Affairs personnel in their function. Usually, a crisis will appear unforeseen or suddenly, or a situation that appeared under control will worsen within a short period of time. Typically, a crisis involves both time pressure and emotional pressure. A quick and well-planned action will be required to improve the situation and/or prevent further damage.
Data Processing	DP
Denomination Commune International (see INN)	DCI
Deutsches Arzneibuch (Germany) (German Pharmacopoeia)	DAB

Document for Regulatory Purposes	The term signifies any document that is intended for regulatory purposes (e.g., application for clinical trial authorization or application for marketing authorization).
Dossier	The term signifies a compilation of documents relevant for a specific regulatory purpose (e.g., application for clinical trial authorization or application for marketing authorization) in a specified country(ies) for a developmental or marketed medicinal product in a structured form (i.e., submission-like). If applicable, it is a subset of the global dossier. The dossier is the basis for the submission(s).
Drug Application Methodology with Optical Storage	DAMOS
Drug Information Association	DIA
Drug Master File	DMF
Education	The term signifies both internally and externally organized, theoretical, and/or practical measures to ensure achievement and/or maintenance of a high standard of knowledge, skills, experience required within Regulatory Affairs.
Electronic Data Processing	EDP
Electronic Standards for the Transmission of Regulatory Information	ESTRI
Electronic Submission	The term signifies a submission that contains some or all of its information in an electronic format.
Employee	The term means a person permanently employed by Regulatory Affairs, except if otherwise specified.

Environmental Protection	The term signifies the responsible use of energy and material within Regulatory Affairs.
Environmental Protection Agency (U.S.)	EPA
Establishment License Application (U.S.)	ELA
Ethical Products	EP
Europaeische Gemeinschaft (Germany) (old for European Community)	EG
European Agency for the Evaluation of Medicinal Products (EU)	EMEA
European Community (old)	EC
European Economic Area	EEA
European Federation of Pharmaceutical Industries' Associations	EFPIA
European Free Trade Association	EFTA
European Pharmacopoeia	EP
European Public Assessment Report	EPAR
European Society of Regulatory Affairs	ESRA
European Union	EU
Federal Food, Drug, and Cosmetic Act	FD&C Act
First Assessment of Efficacy	FAE
Food and Drug Administration (U.S.)	FDA
Freedom of Information (U.S.)	FOI
Free Sales Certificate	FSC

Gesetz zur Neurordnung des AMG (Germany) (Drug Law as amended)	AMNG
Global Dossier	The term signifies a compilation of all documents required for international regulatory purpose(s) for a developmental or already marketed medicinal product. It is maintained continuously throughout the life cycle of the medicinal product and serves as a repository for the generation of dossiers and submissions.
Good Clinical Practice	GCP
Good Clinical Trial Practice	GCTP
Good Drug Regulatory Practice	GDRP
Good Laboratory Practice	GLP
Good Manufacturing Practice	GMP
Good Regulatory Practice	GRP
Guideline	Worldwide guidelines, regulations, laws, relevant publications, position papers, and company experience (e.g., contact reports, deficiency letters, Regulatory Affairs know-how) that may impact the company's substances and medicinal products (e.g., medicinal product development, marketing authorization, and surveillance programs).
Health Maintenance Organization	HMO
Hypertext Markup Language	HTML
Information	The term signifies any knowledge on the company's substances and/or medicinal products that might be relevant for partners within Regulatory Affairs or other contact partners as defined by legal and/or business

	obligations. This covers information received directly, by phone, E-mail, fax, letter, or other route.
Information Technology	IT
Inspection	The term signifies inspections by Regulatory Bodies with regard to Regulatory Affairs (e.g., inspection before issuing a marketing authorization).
Internal Company Standard for Documents for Regulatory Purposes	The term signifies a generic (not product-specific) document that gives the company's evaluation and summary of actual regulatory requirements for the format and content of documents for regulatory purposes, and cross-references regulations and guidelines.
International Classification of Diseases	ICD
International Conference on Harmonisation (EU, Japan, U.S.)	ICH
International Federation of Pharmaceutical Manufacturers Association	IFPMA
International Nonproprietary Name	INN
Investigational New Drug Application (U.S.)	IND
Japanese Adverse Reaction Terminology	JART
Labeling	Labeling reflects the company's actual state of knowledge on its medicinal products, by documenting per medicinal product, scientifically relevant and essential statements. This labeling is the basis for patient leaflet(s), professional information, and promotional material(s).

Letter of Authorization	LOA
Market Authorization by Network Submission and Evaluation	MANSEV
Marketing Authorization	The term signifies an approval by a Regulatory Body with regard to the placing of a medicinal product on the market. MA
Marketing Authorization Application	The term signifies a notification of application to a Regulatory Body with the purpose of placing a medicinal product on the market. MAA
Medical Dictionary for Drug Regulatory Affairs	MEDDRA
Medicinal Product	The term signifies both development/research products and marketed medicinal products throughout the entire life cycle for which the company has legal responsibility, be it as the contract manufacturer, the person responsible for bringing the product into the market, the partner in co-marketing, the receiver or giver of a license, Drug Master File holder or letter of authorization holder, or a partner in joint venture development.
Ministry of Health and Welfare (Japan)	MHW
Multiagency Electronic Regulatory Submission	MERS
New Chemical Entity	NCE
New Drug Application (U.S.)	NDA
New Drug Submission (Canada)	NDS
New Molecular Entity	NME

Outsourcing	In-sourcing, co-sourcing, or outsourcing of Regulatory Affairs work to external parties.
Over-the-counter	OTC
Patient Information Leaflet	PIL
Periodic Safety Update Report	The term signifies a document for regulatory purposes issued by Drug Safety and Regulatory Affairs for the company's marketed medicinal products for safety updates or pharmacovigilance according to regulatory requirements. It is part of, if applicable, the global dossier, and, if nationally required, part of dossier(s) and submission(s).
Pharmaceutical Inspection Convention	PIC
Pharmacodynamics	PD
Pharmacokinetics	PK
Plant Master File	PMF
Policy	Policies define the basic principles under which Regulatory Affairs is to operate worldwide. P
Postmarketing Surveillance	PMS
Prescription only medicinal product	POM product
Product (*see* Medicinal Product)	Product
Product License (UK)	PL
Product License Application (U.S.)	PLA
Promotion/Advertising	The term signifies any published information on the company's substances or medicinal products that is published with a view of making the

	product known and/or augmenting sales (e.g., newspaper ads, television commercials). It excludes patient information and physician's information.
Quality Assurance	QA
Quality Control	QC
Quality Management	QM
Rapporteur (EU)	RAP
Record	The term includes written documents and records maintained on microfilm, optical disc, magnetic tape, or other media. The term includes, if not otherwise specified, not only company records that are kept in official archives or centralized locations but also those company-related records kept in individual files. This also includes records on development/research projects.
Reference Member State (EU)	RMS
Regulatory Affairs	RA
Regulatory Affairs Professional Society	RAPS
Regulatory Strategy	The term signifies the selection of the appropriate submission strategy in terms of the content and presentation of dossier(s), as well as time point(s) for submission(s), and procedure(s) to be used, and considering the target Summary of Product Characteristics for the developmental or marketed medicinal product and the regulatory environment.
Remote Data Entry	RDE

Scale-up and Post-Approval Changes (U.S.)	SUPAC
Soumission Electronique des Dossiers d'Authorisation de Mise sur le Marché (France)	SEDAMM
Standard	These are definitions of items that are required to be the same throughout the organization. S
Standard General Markup Language	SGML
Standard Operating Procedure	This term defines how Policies are implemented and Standards are met in daily operations. SOP
Submission	The term signifies a country-specific compilation of documents for a specific regulatory purpose (e.g., application for clinical trial authorization or application for marketing authorization) for a developmental or marketed medicinal product in a structured form according to national regulatory requirements. It is based on the dossier, if applicable, the global dossier. It may contain additional national documents (e.g., national leaflets or application forms).
Submission Management and Review Tracking (U.S.)	SMART
Substance	The term signifies both development/research products and marketed medicinal products throughout the entire life cycle for which the company has legal responsibility, be it as the contract manufacturer, the person responsible for bringing the product

	into the market, the partner in co-marketing, the receiver or giver of a license, the Drug Master File holder or letter of authorization holder, or a partner in joint venture development.
Summary Basis of Approval	SBA
Summary of Product Characteristics (new)	SMPC
Summary of Product Characteristics (old, *see* SMPC)	SPC
Summary of Product Information	SPI
Supplementary New Drug Application (U.S.)	SNDA
Supplementary New Drug Submission (Canada)	SNDS
Supplementary Protection Certificate	SPC
Table of Contents	TOC
Terminology	The term signifies expressions (including abbreviations, if applicable) frequently used within Regulatory Affairs and/or requiring definition to clarify and standardize the meaning.
Tool	This term signifies all programs or databases designed/purchased for the purpose of facilitating specific functions of Regulatory Affairs.
Training (*see* Education)	
United States Adopted Name	USAN
United States Pharmcopoeia	USP
Variation (*see* Change)	Variation
WHO Adverse Reaction Terminology	WHOART
World Health Organization	WHO

Glossary

Abbreviated New Drug Application/Submission
ANDA (U.S.), ANDS (Canada)

ADE
Adverse Drug Event

ADR
adverse drug reaction

ADROIT
Adverse Drug Reaction On-line Information Trackingsystem (UK)

ADROIT Electronically Generated Information Service
AEGIS

Adverse Drug Event
ADE

Adverse Drug Reaction
ADR

Adverse Drug Reaction On-line Information Trackingsystem
ADROIT (UK)

Adverse Reaction System
AERS (U.S.)

Advertising
see Promotion

AEGIS
ADROIT Electronically Generated Information Service

AERS
Adverse Reaction System (U.S.)

AMG
Arzneimittelgesetz (Drug Law) (Germany)

AMM
Autorisation de Mise sur Marché (marketing authorization)
(France)

AMNG
Gesetz zur Neuordnung des AMG (Drug Law as amended)
(Germany)

Analysis of variance
ANOVA

Anatomical Therapeutic Chemical Classification Index
ATC Code

ANDA
Abbreviated New Drug Application (U.S.)

ANDS
Abbreviated New Drug Submission (Canada)

ANOVA
Analysis of variance

Application for a Clinical Trial License
A notification or application to a Regulatory Body with the purpose of starting local trials in humans.

Application for Marketing Authorization
see Marketing Authorization Application

Archiving Management
The identification, retention, storage, protection, and disposal of
the records of Regulatory Affairs, including records on the developmental research of medicinal products and already marketed
medicinal products.

Arzneimittelgesetz
AMG (Drug Law) (Germany)

ATC Code
Anatomical Therapeutic Chemical Classification Index

Autorisation de Mise Sur Marché
AMM (marketing authorization) (France)

BAN
British Approved Name

BIRA
British Institute of Regulatory Affairs

BP
British Pharmacopoeia

British Approved Name
BAN

British Institute of Regulatory Affairs
BIRA

British Pharmacopoeia
BP

CAMA
Computer-Assisted Marketing Authorization

CANDA
Computer-Assisted New Drug Application

CAPLA
Computer-Assisted Product Licence Application

CARN
Chemical Abstracts Registry Number

CARS
Computer-Assisted Regulatory Submission

Case Record Form
CRF

CBER
Center for Biologics Evaluation and Research (U.S.)

CDER
Center for Drug Evaluation and Research (U.S.)

CEN
Committée Européen de Normalisation

Center for Biologics Evaluation and Research
CBER (U.S.)

Center for Drug Evaluation and Research
CDER (U.S)

Central European Society of Regulatory Affairs
MEGRA

Central Pharmaceutical Affairs Council
CPAC (Japan)

CFR
Code of Federal Regulations (U.S.)

Change
Any changes or variations concerning the company's medicinal products worldwide. This includes, for example, changes in starting materials, manufacturing process/equipment/batch size/site, dosage form, packaging, controls, patient leaflet, labeling, or outer packaging. If subject to submission to the Regulatory Bodies, any additional information, such as advertising/promotional material, is also included.

Chemical Abstracts Registry Number
CARN

Chemistry, Manufacturing, and Controls
CMC

CIOMS
Council for the International Organization of Medicinal Science

Clinical Quality Assurance
CQA

Clinical Trial Application
CTA (UK)

Clinical Trial Certificate
CTC (UK)

Clinical Trial Exemption Application
CTX (UK)

CMC
Chemistry, Manufacturing, and Controls

Code of Federal Regulations
CFR (U.S.)

Coding Symbols for a Thesaurus of Adverse Reaction Terms
COSTART

Committée Européen de Normalisation
CEN

Committee for Proprietary Medicinal Products
CPMP

Committee for Veterinary Medicinal Products
CVMP

Company Comment
A document communicating to Regulatory Bodies comments—suggestions for changing a guideline. Typically, it will be in letter format and contain comments, suggestions, and critical issues, with references to the original document. It may be submitted via industry associations and/or Regulatory Affairs professional societies or directly to the Regulatory Bodies.

Company Position
For each important guideline the result of interpretation/discussion by Regulatory Affairs, and, if applicable, other concerned departments/disciplines. It will contain a summary and an evaluation of, for example, critical issues, possible consequences for the company, and recommendations.

Computer-Assisted Marketing Authorization
CAMA

Computer-Assisted New Drug Application
CANDA

Computer-Assisted Product License Application
CAPLA

Computer-Assisted Regulatory Submission
CARS

Contact Partner
Person(s), department(s), company(ies), and/or Regulatory Body(ies) with which Regulatory Affairs personnel interact in a business environment.

Contact Report
A report on contact with Regulatory Bodies in a standardized format.

Contact with Regulatory Body
Any communication between Regulatory Affairs and a Regulatory Body directly, by phone, E-mail, fax, telex, letter, or during a personal meeting.

Contract Research Organization
CRO

COSTART
Coding Symbols for a Thesaurus of Adverse Reaction Terms

Council for the International Organization of Medicinal Science
CIOMS

CPAC
Central Pharmaceutical Affairs Council (Japan)

CPMP
Committee for Proprietary Medicinal Products

CQA
Clinical Quality Assurance

CRF
Case Record Form

Crisis

Any situation with a high potential of danger or damage to the company's reputation, substances, or medicinal products, or personnel within the scope of Regulatory Affairs's responsibilities or closely connected with its function or directly affecting Regulatory Affairs's personnel in their function. Usually, a crisis will appear unforeseen or suddenly, or a situation that appeared under control will worsen within a short period of time. Typically, a crisis involves both time pressure and emotional pressure. A quick and well-planned action will be required to improve the situation and/or prevent further damage.

CRO

Contract Research Organization

CTA

Clinical Trial Application (UK)

CTC

Clinical Trial Certificate (UK)

CTX

Clinical Trial Exemption Application (UK)

CVMP

Committee for Veterinary Medicinal Products

DAB

Deutsches Arzneibuch (German Pharmacopoeia) (Germany)

DAMOS

Drug Application Methodology with Optical Storage

Data Processing

DP

DCI

Dénomination Commune International (*see* INN)

Dénomination Commune International

DCI (*see* INN)

Deutsches Arzneibuch

DAB (German Pharmacopoeia) (Germany)

DIA
Drug Information Association

DMF
Drug Master File

Document for Regulatory Purposes
Any document that is intended for regulatory purposes (e.g., application for clinical trial authorization or application for marketing authorization).

Dossier
A compilation of documents relevant for a specific regulatory purpose (e.g., application for clinical trial authorization or application for marketing authorization) in a specified country(ies) for a developmental or marketed medicinal product in a structured form (i.e., submission-like). If applicable, it is a subset of the global dossier. The dossier is the basis for the submission(s).

DP
Data Processing

Drug Application Methodology with Optical Storage
DAMOS

Drug Information Association
DIA

Drug Master File
DMF

EC
European Community (old)

ECPHIN
Electronic Communication of Pharmaceutical Information (EU)

EDP
Electronic Data Processing

Education
Internally and externally organized, theoretical, and/or practical measures to ensure achievement and/or maintenance of a high

standard of knowledge, skills, and experience required within Regulatory Affairs.

EEA
European Economic Area

EFPIA
European Federation of Pharmaceutical Industries' Associations

EFTA
European Free Trade Association

ELA
Establishment License Application (U.S.)

Electronic Communication of Pharmaceutical Information
ECPHIN (EU)

Electronic Data Processing
EDP

Electronic Standards for the Transmission of Regulatory Information
ESTRI

Electronic Submission
A submission that contains some or all of its information in an electronic format.

EMEA
European Agency for the Evaluation of Medicinal Products (EU)

Employee
A person permanently employed by Regulatory Affairs, except if otherwise specified.

Environmental Protection
The responsible use of energy and material within Regulatory Affairs.

Environmental Protection Agency
EPA (U.S.)

EP
Ethical Products

EP
European Pharmacopoeia

EPA
Environmental Protection Agency (U.S.)

EPAR
European Public Assessment Report

ESRA
European Society of Regulatory Affairs

Establishment License Application
ELA (U.S.)

ESTRI
Electronic Standards for the Transmission of Regulatory Information

Ethical Products
EP

EU
European Union

European Agency for the Evaluation of Medicinal Products
EMEA (EU)

European Community
EC (old)

European Economic Area
EEA

**European Federation of Pharmaceutical Industries'
Associations**
EFPIA

European Free Trade Association
EFTA

European Pharmacopoeia
EP

European Public Assessment Report
EPAR

European Society of Regulatory Affairs
ESRA

European Union
EU

FAE
First Assessment of Efficacy

FDA
Food and Drug Administration (U.S.)

FD&C Act
Federal Food, Drug, and Cosmetics Act

Federal Food, Drug, and Cosmetics Act
FD&C Act

First Assessment of Efficacy
FAE

FOI
Freedom of Information (U.S.)

Food and Drug Administration
FDA (U.S.)

Freedom of Information
FOI (U.S.)

Free Sales Certificate
FSC

FSC
Free Sales Certificate

GCP
Good Clinical Practice

GCTP
Good Clinical Trial Practice

GDRP
Good Drug Regulatory Practice

Gesetz zur Neuordnung des AMG
AMNG (Drug Law as amended) (Germany)

Global Dossier
A compilation of all documents required for international regulatory purpose(s) for a developmental or already marketed medicinal product. It is maintained continuously throughout the life cycle of the medicinal product and serves as a repository for the generation of dossiers and submissions.

GLP
Good Laboratory Practice

GMP
Good Manufacturing Practice

Good Clinical Practice
GCP

Good Clinical Trial Practice
GCTP

Good Drug Regulatory Practice
GDRP

Good Laboratory Practice
GLP

Good Manufacturing Practice
GMP

Good Regulatory Practice
GRP

GRP
Good Regulatory Practice

Guideline
Worldwide guidelines, regulations, laws, relevant publications, position papers, and company experience (e.g., contact reports, deficiency letters, Regulatory Affairs know-how) that may impact the company's substances and medicinal products (e.g., medicinal product development, marketing authorization, and surveillance programs).

Health Maintenance Organization
HMO

HMO
Health Maintenance Organization

HTML
Hypertext Markup Language

Hypertext Markup Language
HTML

ICD
International Classification of Diseases

ICH
International Conference on Harmonisation (EU, Japan, U.S.)

IFPMA
International Federation of Pharmaceutical Manufacturers Association

IND
Investigational New Drug Application (U.S.)

Information
Any knowledge on the company's substances and/or medicinal products that might be relevant for partners within Regulatory Affairs or other contact partners as defined by legal and/or business obligations. This covers information received directly, by phone, E-mail, fax, letter or other route.

Information Technology
IT

INN
International Nonproprietary Name

Inspection
Inspections by Regulatory Bodies with regard to Regulatory Affairs (e.g., inspection before issuing a marketing authorization).

INTDIS
International database accessible via SWEDIS

Internal Company Standard for Documents for Regulatory Purposes
A generic (not product-specific) document that gives the company's evaluation and summary of actual regulatory requirements for the format and content of documents for regulatory purposes, and cross-references regulations and guidelines.

International Classification of Diseases
ICD

International Conference on Harmonisation
ICH (EU, Japan, U.S.)

International database accessible via SWEDIS
INTDIS

International Federation of Pharmaceutical Manufacturers Association
IFPMA

International Nonproprietary Name
INN

Investigational New Drug Application
IND (U.S.)

IT
Information Technology

Japanese Adverse Reaction Terminology
JART (Japan)

JART
Japanese Adverse Reaction Terminology (Japan)

Labeling
Labeling reflects the company's actual state of knowledge on its medicinal products, by documenting per medicinal product the scientifically relevant and essential statements. This labeling is the basis for patient leaflet(s), professional information, and promotional material(s).

Letter of Authorization
LOA

LOA
Letter of Authorization

MA
Marketing Authorization

MAA
Marketing Authorization Application

MANSEV
Market Authorization by Network Submission and Evaluation

Market Authorization by Network Submission and Evaluation
MANSEV

Marketing Authorization (MA)
An approval by a Regulatory Body with regard to the placing of a medicinal product on the market.

Marketing Authorization Application (MAA)
A notification of an application to a Regulatory Body with the purpose of placing a medicinal product on the market.

MEDDRA
Medical Dictionary for Drug Regulatory Affairs

Medical Dictionary for Drug Regulatory Affairs
MEDDRA

Medicinal Product
Both developmental/research products and marketed medicinal products throughout the entire life cycle for which the company has legal responsibility, be it as the contract manufacturer, the person responsible for bringing the product into the market, the partner in co-marketing, the receiver or giver of a license, Drug Master File (DMF) holder or letter of authorization holder, or a partner in joint venture development.

MEGRA
Central European Society of Regulatory Affairs

MERS
Multiagency Electronic Regulatory Submission

MHW
Ministry of Health and Welfare (Japan)

Ministry of Health and Welfare
MHW (Japan)

Multiagency Electronic Regulatory Submission
MERS

NCE
New Chemical Entity

NDA
New Drug Application (U.S.)

NDS
New Drug Submission (Canada)

New Chemical Entity
NCE

New Drug Application
NDA (U.S.)

New Drug Submission
NDS (Canada)

New Molecular Entity
NME

NME
New Molecular Entity

OTC
Over-the-counter

Outsourcing
In-sourcing, co-sourcing, or outsourcing of Regulatory Affairs work to external parties.

Over-the-counter
OTC

P
Policy

Patient Information Leaflet
PIL

PD
Pharmacodynamics

Periodic Safety Update Report (PSUR)
A document for regulatory purposes issued by Drug Safety and Regulatory Affairs for the company's marketed medicinal products for safety updates or pharmacovigilance according to regulatory requirements. It is part of, if applicable, the global dossier, and, if nationally required, part of dossier(s) and submission(s).

Pharmaceutical Inspection Convention
PIC

Pharmacodynamics
PD

Pharmacokinetics
PK

PIC
Pharmaceutical Inspection Convention

PIL
Patient Information Leaflet

PK
Pharmacokinetics

PL
Product License (UK)

PLA
Product License Application (U.S.)

Plant Master File
PMF

PMF
Plant Master File

PMS
Postmarketing Surveillance

Policy (P)
Policies define the basic principles under which Regulatory Affairs is to operate worldwide.

POM
Prescription only medicinal product

Postmarketing Surveillance
PMS

Prescription only medicinal product
POM

Product
see Medicinal Product

Product License
PL (UK)

Product License Application
PLA (U.S.)

Promotion/Advertising
Any published information on the company's substances or medicinal products that is published with a view of making the product known and/or augmenting sales (e.g., newspaper ads, television commercials). It excludes patient information and physician's information.

PSUR
Periodic Safety Update Report

QA
Quality Assurance

QC
Quality Control

QM
Quality Management

Quality Assurance
QA

Quality Control
QC

Quality Management
QM

RA
Regulatory Affairs

RAP
Rapporteur (EU)

Rapporteur
RAP (EU)

RAPS
Regulatory Affairs Professionals Society

RDE
Remote Data Entry

Record
Written documents and records maintained on microfilm, optical disc, magnetic tape, or other media. The term includes, if not otherwise specified, not only company records that are kept in official archives or centralized locations but also those company-related records kept in individual files. This also includes records on developmental/research projects.

Reference Member State
RMS (EU)

Regulatory Affairs
RA

Regulatory Affairs Professionals Society
RAPS

Regulatory Body
Any Regulatory Body authorized to carry out inspection(s), or to issue clinical trial authorization(s) or marketing authorization(s), etc.

Regulatory Strategy
The selection of the appropriate submission strategy in terms of the content and presentation of dossier(s), as well as time point(s) for submission(s), and procedure(s) to be used, and considering the target Summary of Product Characteristics for the developmental or marketed medicinal product and the regulatory environment.

Remote Data Entry
RDE

RMS
Reference Member State (EU)

S
Standard

SBA
Summary Basis of Approval

Scale-up and Post-Approval Changes
SUPAC (U.S.)

SEDAMM
Soumission Electronique des Dossiers d'Autorisation de Mise Sur le Marché (France)

SGML
Standard General Markup Language

SMART
Submission Management and Review Tracking (U.S.)

SMPC
Summary of Product Characteristics (new)

SNDA
Supplementary New Drug Application (U.S.)

SNDS
Supplementary New Drug Submission (Canada)

SOP
Standard Operating Procedure

Soumission Electronique des Dossiers d'Authorisation de Mise sur le Marché
SEDAMM

SPC
Summary of Product Characteristics (old); *see* SMPC

SPC
Supplementary Protection Certificate

SPI
Summary of Product Information

Standard (S)
Standards are definitions of items that are required to be the same throughout the organization.

Standard General Markup Language
SGML

Standard Operating Procedure (SOP)
Standard Operating Procedures define how policies are implemented and standards are met in daily operations.

Submission
A country-specific compilation of documents for a specific regulatory purpose (e.g., application for clinical trial authorization of application for marketing authorization) for a developmental or marketed medicinal product in a structured form according to national regulatory requirements. It is based on the dossier, or, if applicable, the global dossier. It may contain additional national documents (e.g., national leaflets or application forms).

Submission Management and Review Tracking
SMART (U.S.)

Substance
Both developmental/research products and marketed medicinal products throughout the entire life cycle for which the company has legal responsibility, be it as the contract manufacturer, the person responsible for bringing the substance/product into the market, the partner in co-marketing, the receiver or giver of a license, the Drug Master File (DMF) holder or letter of authorization holder, or a partner in joint venture development, etc.

Summary Basis of Approval
SBA

Summary of Product Characteristics
SMPC (new)

Summary of Product Characteristics
SPC (old); *see* SMPC

Summary of Product Information
SPI

SUPAC
Scale-up and Post-approval Changes (U.S.)

Supplementary New Drug Application
SNDA (U.S.)

Supplementary New Drug Submission
SNDS (Canada)

Supplementary Protection Certificate
SPC

Table of Contents
TOC

Terminology
Expressions (including abbreviations, if applicable) frequently used within Regulatory Affairs and/or requiring definition to clarify and standardize the meaning.

TOC
Table of Contents

Tool
All programs or databases designed/purchased for the purpose of facilitating specific functions of Regulatory Affairs.

Training
see Education

United States Adopted Name
USAN

United States Pharmacopeia
USP

USAN
United States Adopted Name

USP
United States Pharmacopoeia

Variation
see Change

WHO
World Health Organization

WHOART
World Health Organization Adverse Reaction Terminology

World Health Organization
WHO

World Health Organization Adverse Reaction Terminology
WHOART

Index

Milton Keynes UK
Ingram Content Group UK Ltd.
UKHW051942071024
449327UK00026B/2136